The Twin Paradox Explained

Other books written by Dave Karpinsky

Artificial Intelligence & Information Technology

- Artificial Intelligence (AI) for Daily Life: A Practical Guide to Artificial Intelligence
- AI and Creativity: How Machines are Changing Art, Music & Literature
- AI-Powered PM: Leveraging Artificial Intelligence for Enhanced Efficiency and Success
- Artificial Intelligence Rise and Humanity Fall
- Data-Driven Future: Harnessing AI and Big Data for Tomorrow's Challenges
- Deepfake Technology: The Dark Side of AI, Manipulation and Digital Deception.
- Fixing Failed Projects: How to Master the Art of Project Turnaround
- From Data to Decisions: The Role of AI in Business Intelligence
- Jobs AI Will Replace: Re-tool or Be Left Behind
- Mastering Advanced Project Management: Strategies for Excellence
- Mastering Project Management: In complex, stressful & high-pressure environments
- SAP S/4 Implementation: A Comprehensive Guide for Practitioners
- SAP S/4 Implementation Methodologies
- SAP S/4 Implementation – Volume 1: Prep & Explore Phases
- SAP S/4 Implementation – Volume 2: Realize & Deploy Phases
- SAP S/4 Implementation – Volume 3: When Projects Fail
- The Five-Day Organizational Change Manager
- The Five--Day Project Manager
- The Project Management Masterclass: Advanced Techniques for Success

- The Rise of Real-Time Analytics: Speed, Precision, and Competitive Edge

Business & Finance
- Building Wealth in Developing Nations: A Comprehensive Step-by-Step Guide to Empower Emerging Markets
- Chief Executive's (CxO) Playbook: The First 90 Days Guide to Success
- Creating a Deployment Plan: Navigating Complexity to Deliver Success
- Creating a Strategic Roadmap: Crafting the Blueprint from Vision to Execution
- Investing Strategies of the Rich and Famous: Discover How to Diversify Your Portfolio for Maximum Returns
- Outsmart the Game: Winning When the Rules Are Rigged
- The Data Delusion: Exposing False Metrics That Shape Your World
- Trust is the New Currency: How Connection Wins in the Age of AI

Life Coach & Mentor Series
- Aspiring Entrepreneurs
- Bored Housewife
- Career Transition
- Couples and Relationships
- Mid-Life Crisis
- Mindful Healthy Living
- Project Managers
- Seeking Life's Purpose
- Surviving Holidays with In-laws

Science & Physics

- Game Over. Reset Earth
- Quantum Entanglement: The God Effect and the Secrets of Reality
- Multiverse Parallel Dimensions: The Theories and Possibilities of Parallel Universes
- Space-Time Continuum: Navigating the Quantum of the Fourth Dimension
- The Hubble Tension: The Universe's Expansion, Cosmology Crisis, and the Limits of the Big Bang Theory
- The Singularity Shift: Unveiling the Future of Humanity and Intelligence
- Twin Paradox: Solving the Puzzle of Special Relativity

Sociology & Politics

- America at War: Russia, China, Iran, S Korea
- Blue Zones Volume 1: Mystery and Science of Blue Zones
- Blue Zones Volume 2: Longevity Lessons of Blue Zones
- Decline of American Supremacy: Understanding the Erosion, Shaping the Future
- Future of Military Technology Powered by AI: How countries are transforming their warfare
- Herd Instinct: Understanding the Human Psychology of Collective Behavior
- Our Idiot Species: Evolution in Reverse
- Preventing Squatters: A Comprehensive Guide to Protecting Your Property
- Puppet Masters: The Hidden Hands of Political Power
- The Great War of China vs Russia: A Future Battlefield that Reshapes the World

- The Modern Stoic: 365 Ancient Practices for Wisdom, Peace, Purpose ad Strength
- The Next Battlefield: How AI, Robotics, and Biotechnology are Transforming Warfare
- The Savage Guide to Winning: The Brutal Truth About Success
- The Trump Effect: Return to the White House
- The Vatican Murder Cover-Up
- Unf*k Yourself: A No-Bullsh!t Guide to Taking Control
- Warfare Redefined: Military Technologies and Tactics of Tomorrow's Superpowers
- Zero F*cks Given: How to Stop Worrying and Live Your Life
- God & AI Series:
  - Is There God: According to Artificial Intelligence (AI)
  - What is God: According to Artificial Intelligence (AI)
  - What is God's Plan: According to Artificial Intelligence (AI)

"In the world of special relativity, the fastest way to change your future is to move through space at near-light speed."
— *Dave Karpinsky*

The
Twin Paradox
Explained:

Solving the Puzzle of
Special Relativity

Dave Karpinsky, PhD, MBA, PMP, Prosci

Green Parrot Media

Contents

Preface 14

1: The Genesis of Special Relativity 17

2: The Fundamentals of Special Relativity 25

3: Introducing the Twin Paradox 33

4: The Mechanics of Time and Space-Time 41

5: Analyzing the Twin Paradox 53

6: Historical Perspectives and Theoretical
Interpretations 63

7: Experimental Evidence Supporting Relativity 73

8: The Twin Paradox in Popular Culture 83

9: Extensions and Connections to Other Theories 91

10: Conclusion and Reflections 101

Glossary 109

References 115

Preface

The Twin Paradox has fascinated and perplexed physicists, students, and science enthusiasts for over a century. Albert Einstein introduced the concept as part of his theory of Special Relativity. It challenges our intuitive understanding of time, space, and reality itself. It's a thought experiment that reveals the intricate and non-intuitive nature of time dilation, one of the most profound implications of traveling at relativistic speeds.

When I first encountered the Twin Paradox, I was struck by the elegance of its simplicity and the depth of its implications. Two twins, one journeying through space at near-light speed and the other remaining on Earth — how could such a scenario result in different time experiences for each twin? The question isn't just about physics; it touches on the nature of existence and the fabric of the universe.

This book is born from my journey through the landscapes of Special Relativity, a journey filled with moments of clarity, confusion, and, ultimately, understanding. I aim to provide readers with a clear, comprehensive, and engaging explanation of the Twin Paradox, demystifying the concepts of time dilation and relative simultaneity at its core.

The Twin Paradox is not just a puzzle to be solved; it is a doorway into a deeper understanding of the universe. Through this book, I am here to guide you through this doorway, helping you understand the paradox's resolution and the broader implications of Einstein's groundbreaking theory.

As we embark on this exploration together, I encourage you to approach the material with curiosity and an open mind. The Twin Paradox challenges our perceptions and enriches our understanding of the cosmos. By the end of this journey, I hope you will see the universe—and perhaps even time itself—in a new light.

Thank you for joining me in unraveling one of the most intriguing puzzles of modern physics. I am excited to share this exploration with you and illuminate Special Relativity's wonders in the following pages.

—Dave Karpinsky

"Time doesn't tick the same for everyone — it bends to motion, warps with speed, and rewrites the meaning of 'now' with every cosmic mile."
— *Dave Karpinsky*

1: The Genesis of Special Relativity

The Prelude to Relativity: Historical Context and the Limitations of Classical Physics

In the late 19th century, the scientific community believed it had reached the pinnacle of understanding the natural world. Classical physics, epitomized by Isaac Newton's works, provided a robust framework for explaining the motion of objects, the forces they exerted, and the interactions of matter.

Newton's laws of motion and universal gravitation have stood the test of time, describing the universe's mechanics with astonishing precision. The deterministic nature of these laws suggests a clockwork, predictable, and absolute universe. In this universe, time and space are immutable and uniform, serving as the stage on which the drama of physics plays out.

However, cracks were beginning to appear in the foundations of classical physics. As experimental techniques improved and new phenomena were discovered, it became increasingly clear that Newtonian mechanics could not explain everything. One of the most glaring issues was the behavior of light. The Michelson-Morley experiment of 1887, often hailed as one of the most significant null results in the history of science, was designed to detect the presence of the "aether," a supposed medium through which light waves propagated, much like sound waves travel through air. According to the prevailing theories, the speed of light should vary depending on the Earth's motion through this aether. However, the experiment revealed something startling: the speed of light was constant, regardless of the Earth's motion. This was a direct contradiction to the predictions of classical physics.

At the same time, the study of electromagnetic radiation uncovered other anomalies. James Clerk Maxwell's equations, which unified electricity and magnetism into a single theory of electromagnetism, implied that light was an electromagnetic wave. Yet, these equations also suggested that the speed of light was constant—a result that did not fit neatly within the Newtonian framework. If the speed of light was indeed a constant, then something was amiss in how scientists understood time and space.

These inconsistencies hinted at deeper problems within the foundations of physics. The notion of absolute space and time, so central to Newton's worldview, was being called into question. The universe was beginning to look less like a rigid, deterministic machine and more like a puzzle with missing pieces. Classical physics, while powerful, was not complete. A new framework was needed to reconcile these emerging paradoxes and provide a more comprehensive understanding of the natural world.

Einstein's Revolution: Introduction to Einstein's Theories and the Concept of Relativity

Enter Albert Einstein, a young patent clerk in Bern, Switzerland, with a mind unbound by his time's conventions. In 1905, a year often referred to as his "annus mirabilis" or "miracle year," Einstein published four groundbreaking papers that would change the course of physics forever. Among these was his paper on Special Relativity, a theory that would revolutionize our understanding of time, space, and the very fabric of reality.

Einstein's genius lay in his ability to question the most fundamental assumptions about the universe. Where others saw contradictions, he saw opportunities to rethink the very foundations of physics. He asked a simple yet profound question: What if the speed of light is indeed constant, not just in some exceptional cases, but universally, in all frames of reference? From this single postulate, Einstein began to unravel the mysteries that had confounded physicists for decades.

The theory of Special Relativity was born out of two key ideas. First, the laws of physics are the same in all inertial frames of reference—this means that whether you are standing still, moving at a constant speed, or even traveling at near-light speed, the fundamental laws governing the universe remain unchanged. Second, the speed of light is constant in all these frames of reference. It is independent of the motion of the light source or the observer.

These two principles might seem straightforward, but their implications are profound. If the speed of light is constant, then time and space cannot be absolute as Newton had imagined. Instead, they must be flexible and capable of stretching or contracting depending on the observer's relative motion. This was the birth of the concept of "relativity"—the idea that time and space are not fixed but relative to the observer's frame of reference.

Einstein's theory had immediate and startling consequences. It implied that time could slow down or speed up depending on how fast one moves relative to something else — a phenomenon known as time dilation. It suggested that lengths could contract as objects approached the speed of light, a concept known as length contraction. It introduced the world to mass-energy equivalence, encapsulated in the famous equation $E=mc^2$, which revealed that mass could be converted into energy and vice versa.

The implications of Special Relativity were both thrilling and unsettling. It upended centuries of classical physics and challenged people's thoughts about the universe. Suddenly, time was no longer a universal constant; it could vary from observer to observer. Space was no longer a passive backdrop; it could warp and bend under the influence of motion. The universe, it seemed, was far more mysterious and complex than anyone had imagined.

The Role of Space-Time: How Einstein's Work Transformed the Understanding of Time and Space

One of Einstein's most profound contributions to modern physics was his reimagining of time and space as a single, interconnected entity known as space-time. Before Einstein, time and space were considered separate, independent dimensions, and time ticked away uniformly while space extended infinitely in all directions. But Einstein's theory of relativity revealed that time and space are inextricably linked, forming a four-dimensional continuum that underpins the entire universe.

To understand the concept of space-time, it's helpful to visualize it as a fabric — a smooth, continuous sheet that can be warped and stretched. Like stars and planets, massive objects create indentations or "curves" in this fabric, much like a heavy ball placed on a rubber sheet. These curves in space-time are what we perceive as gravity: the more massive an object, the deeper the curve and the stronger the gravitational pull.

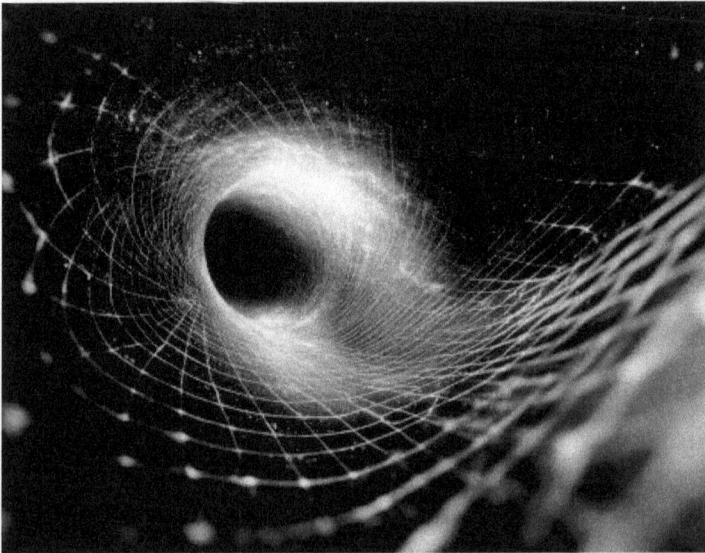

But space-time is not just a static backdrop but dynamic and responsive. The motion of objects through space-time affects the fabric itself, causing it to warp and twist. This warping of space-time leads to time dilation and length contraction. As an object moves faster, it distorts the space-time around it, altering the passage of time and the perception of distance.

One of the most famous illustrations of this concept is the "twin paradox," which we will explore later in this book. The paradox describes a scenario where one twin travels at near-light speed while the other remains on Earth. Upon the traveling twin's return, they find they have aged less than their Earth-bound sibling. This apparent contradiction directly results from how motion affects space-time fabric, causing time to pass differently for each twin.

Einstein's reimagining of time and space had profound implications for our understanding of the universe. It meant that the cosmos was not a static, unchanging entity but a dynamic, interconnected web of relationships between objects, motion, and the very fabric of reality. Space and time were no longer absolutes; they were relative, fluid, and dependent on the observer's frame of reference.

This new understanding of space-time also laid the groundwork for Einstein's later work on General Relativity, which expanded the concepts of Special Relativity to include gravity and acceleration. In General Relativity, gravity is not a force that acts at a distance, as Newton had described, but a manifestation of the curvature of space-time caused by mass and energy. This insight led to some of the most extraordinary predictions in physics, including the existence of black holes, the expansion of the universe, and the bending of light by gravity—predictions that have since been confirmed by observation and experiment.

The journey from Newton's deterministic universe to Einstein's relativistic cosmos was not just a shift in scientific theory; it was a transformation in how humanity perceives the very nature of reality. Einstein's work shattered the comforting certainty of absolute time and space, replacing it with a far more intricate and interconnected universe than we had ever imagined.

As we move forward in this book, we will delve deeper into the implications of Special Relativity, exploring how it challenges our intuitions and reshapes our understanding of the universe. We will examine the twin paradox in detail, unraveling the time dilation and relative simultaneity puzzle. But before we embark on that journey, we must appreciate the historical and conceptual context in which these ideas emerged. Einstein's revolution was not just a leap in scientific thought; it was a profound reimagining of the very fabric of reality—a reimagining that continues to inspire and challenge us today.

2: The Fundamentals of Special Relativity

The Two Postulates: The Constancy of the Speed of Light and the Principle of Relativity

At the heart of Einstein's revolutionary theory of Special Relativity are two deceptively simple yet profoundly transformative postulates. These ideas, proposed in 1905, laid the groundwork for a new understanding of the universe that would challenge the very fabric of reality as it was known at the time.

The first postulate, the **Principle of Relativity**, states that the laws of physics are the same in all inertial frames of reference. This means that no matter how fast you move, the fundamental laws governing physical phenomena remain consistent as long as you are not accelerating. Imagine being on a smoothly flying airplane, with no sense of the plane's movement — objects behave the same way if you were stationary on the ground. This idea was not entirely new; Galileo had proposed a similar principle centuries earlier. However, Einstein extended this concept, asserting that even the laws governing light and electromagnetism are consistent across all inertial frames.

The second postulate is the **Constancy of the Speed of Light**. It asserts that the speed of light in a vacuum is always the same — approximately 299,792 kilometers per second (186,282 miles per second) — regardless of the motion of the light source or the observer. This idea was radical because it directly contradicted the common sense of the time. Traditionally, velocities were thought to add up; if you were moving towards a light source, you would expect to measure the light's speed as more significant than if you were moving away from it. But Einstein's postulate defied this intuition, suggesting that light's speed is a universal constant, unaltered by the observer's relative motion.

These two postulates were like puzzle pieces that, when fit together, revealed a new picture of reality — one where space and time were not as they seemed. The implications were staggering. If the speed of light is constant, and the laws of physics are the same in all inertial frames, then space and time must adjust to preserve this constancy. This realization opened the door to concepts that defy everyday experience: time dilation, length contraction, and the relativity of simultaneity.

These ideas were initially met with skepticism, as they appeared to upend the well-established Newtonian worldview. Yet, as we will see, these postulates resolved the inconsistencies that plagued classical physics and provided a more profound understanding of the universe. In this universe, the passage of time and the dimensions of space are not absolute but relative, dependent on the observer's motion.

Time Dilation: Understanding How Time Slows Down for Moving Objects

Imagine you are aboard a spacecraft hurtling through space at a significant fraction of the speed of light. You look at your wristwatch and see time ticking by as usual.

Yet, to an observer on Earth, your clock appears to tick more slowly. This phenomenon, known as **time dilation**, is one of the most mind-bending consequences of Special Relativity.

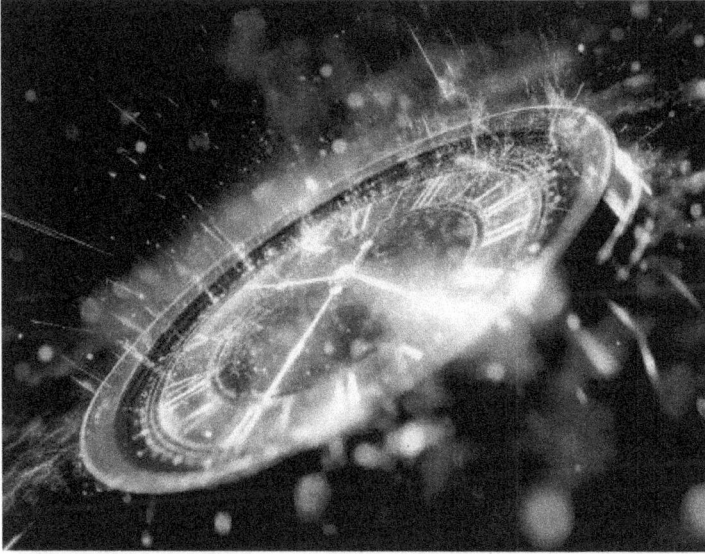

Time dilation arises directly from the two postulates of Special Relativity. If the speed of light is constant in all inertial frames, then time must slow down for a moving observer relative to a stationary one. To grasp this, consider a thought experiment that Einstein might have pondered.

Picture a simple clock that measures time by bouncing a beam of light between two mirrors. In the clock's frame of reference, the light travels straight up and down between the mirrors. Now, imagine this clock is aboard a fast-moving spaceship. To an observer on Earth, the light beam follows a more extended, diagonal path because the spacecraft is moving forward as the light bounces between the mirrors. Since the speed of light is constant, the longer path means that, from the Earth observer's perspective, the light takes more time to complete one bounce. Therefore, the clock appears to tick more slowly. The faster the spaceship moves, the greater the time dilation.

This effect is not just a theoretical curiosity; numerous experiments have confirmed it. For instance, highly accurate atomic clocks flown worldwide on jets or placed in fast-moving spacecraft have been shown to tick more slowly than identical clocks on the ground. The implications of time dilation are profound, especially in the context of the Twin Paradox, where one twin ages slower than the other due to their high-speed journey through space.

Time dilation also has practical implications in our modern world. The Global Positioning System (GPS), which many rely on daily, must account for time dilation. Satellites in orbit around the Earth move at high speeds relative to the surface, and their clocks tick slightly slower than those on the ground. Engineers must correct for this time dilation to ensure the accuracy of GPS signals. Without these relativistic corrections, the system would quickly become inaccurate, leading to errors in navigation.

Time dilation forces us to rethink our understanding of time. It reveals that time is not an absolute, universal quantity but a relative one, dependent on the observer's state of motion. This counterintuitive realization challenges the common-sense notion that time flows uniformly everywhere. Yet, as unsettling as it may be, time dilation is a fundamental aspect of our universe woven into the very fabric of reality.

Length Contraction: How Objects Contract in the Direction of Motion at High Speeds

If time can slow down, what happens to space? Special Relativity tells us that just as time dilates, lengths contract along the direction of motion — a phenomenon known as **length contraction**. This idea is as strange as time dilation, yet it is a necessary consequence of Einstein's postulates.

To visualize length contraction, let's return to our fast-moving spaceship. Suppose the spacecraft is traveling close to the speed of light, and an observer on Earth measures the spaceship's length. To the observer, the spacecraft would appear shorter along the direction of its motion than at rest. The faster the spacecraft moves, the more pronounced the contraction becomes. At the speed of light, in theory, the spaceship's length in the direction of motion would shrink to zero—a concept that defies everyday experience.

The mathematical expression for length contraction is given by the Lorentz factor, which also governs time dilation. The Lorentz factor, denoted by the Greek letter gamma (γ), depends on the relative velocity between the observer and the moving object. As the object's speed approaches the speed of light, the Lorentz factor increases, leading to a more significant contraction.

Length contraction, like time dilation, has been confirmed through experiments. One of the most famous involves high-speed particles, such as muons, created in the Earth's upper atmosphere by cosmic rays. These particles travel toward the Earth's surface at speeds close to the speed of light. Despite their short lifespans, muons are observed to reach the ground, which should be impossible given their expected decay times. The explanation lies in length contraction. From the muon's perspective, the distance to the Earth's surface is contracted, allowing it to cover the distance before decaying.

Length contraction has implications for our understanding of space itself. It suggests that space is not a rigid, unchanging entity but something that can be stretched or compressed depending on the observer's motion. This idea was revolutionary, challenging the Newtonian concept of absolute space. In Einstein's universe, space, like time, is relative.

The concept of length contraction also offers a deeper insight into the Twin Paradox. When one twin embarks on a high-speed journey, not only does time slow down for them, but the distances they travel also contract. This means that while the traveling twin experiences less time, they also perceive the distance they travel as shorter than it would appear to their stationary sibling. Both time dilation and length contraction work together to create the seemingly paradoxical situation where the traveling twin returns younger.

Length contraction, like time dilation, is not something we notice in everyday life because the effects only become significant at speeds close to that of light. However, in high-energy physics, where particles routinely move at relativistic speeds, these effects are crucial for understanding the universe's behavior on the most minor scales.

Conclusion

The fundamentals of Special Relativity — the constancy of the speed of light, time dilation, and length contraction — offer a glimpse into our universe's profound and often counterintuitive nature. These concepts challenge the Newtonian worldview that had dominated physics for centuries, replacing it with a reality where time and space are not absolute but relative to the observer's motion.

As we continue our exploration of the Twin Paradox, we will see how these fundamental principles come together to create one of the most fascinating puzzles in modern physics. The paradox, far from being a mere thought experiment, provides a window into the strange yet beautiful nature of time and space. It shows us that the universe is far more complex than our everyday experiences suggest and that the true nature of reality lies in the interplay between motion, time, and space.

In the next chapter, we will delve deeper into the Twin Paradox itself, unraveling why the traveling twin ages more slowly and what this means for our understanding of the universe. We will explore the implications of relativity for concepts of simultaneity and causality and how the paradox challenges our most fundamental assumptions about time. The journey promises to be as enlightening as it is intriguing, revealing the subtle and elegant workings of Einstein's universe.

3: Introducing the Twin Paradox

The Classic Scenario: Description of the Twin Experiment Involving Space Travel and Time Differences

The Twin Paradox is one of the most famous thought experiments in modern physics, a scenario that perfectly encapsulates the strange and counterintuitive nature of Einstein's theory of Special Relativity. It's a paradox that has intrigued, puzzled, and even bewildered those encountering it for the first time. Still, it is also a key to understanding the more profound implications of relativity.

Imagine two identical twins, Alice and Mary. They were born simultaneously and have lived their lives side by side, experiencing everything together — until one day, a space mission offers an extraordinary opportunity. Alice is selected to embark on a journey to a distant star system, traveling close to the speed of light while Mary stays on Earth. Alice's spacecraft is a marvel of futuristic technology, designed to accelerate to an astonishing fraction of light speed, making the trip possible within her lifetime.

As Alice's spacecraft accelerates away from Earth, she and Mary experience the same passage of time, or so it seems. Alice appears to have only taken a few years to journey to the distant star. She eats, sleeps, and lives aboard the ship, all while time seems to flow as usual. When she finally reaches the star system, she performs her scientific duties and prepares for the return journey, which takes an equivalent amount of time.

Meanwhile, Mary remains on Earth, living her life as usual. She watches the news, goes to work, and marks the days on the calendar, all the while thinking of her sister exploring the cosmos. But when Alice finally returns to Earth and steps off her spacecraft, she is shocked to find that while only a few years have passed for her, decades have passed for Mary. He is now an older woman, while she is still as young as she was when she left. The twins, who were once the same age, are now separated by a vast chasm of time. This is the essence of the Twin Paradox.

This paradox is not just a science fiction scenario — it is a direct consequence of the principles of Special Relativity, particularly the concept of time dilation. The faster an object moves relative to another, the more time slows down for the object in motion. For Alice, traveling at nearly the speed of light, time on board her spacecraft slowed dramatically compared to time on Earth. As a result, she aged much more slowly than her twin brother.

The Twin Paradox is compelling not only because it challenges our everyday understanding of time but also because it forces us to confront the reality that time is not an absolute, universal quantity. Instead, time is relative, dependent on the observer's state of motion. The idea that two people who were once the same age could end up aging at drastically different rates simply because one of them traveled through space at high speed is nothing short of mind-bending.

Initial Confusion: Common Misconceptions and Initial Misunderstandings About the Paradox

When people first encounter the Twin Paradox, it's natural to feel confused or even in disbelief. The scenario seems to defy common sense — how can one twin age more slowly than the other simply by traveling at high speed? The initial reaction often involves a series of misconceptions and misunderstandings, many of which stem from our deeply ingrained notions of time and space as absolute and universal.

One of the most common misconceptions is that the paradox violates the principle of relativity. After all, if the laws of physics are the same in all inertial frames, why should one twin's time experience be different from the other's? Some might argue that from Alice's perspective, it is Mary who is moving away at high speed, and thus, it should be Mary who ages more slowly. This symmetrical view suggests that both twins should experience the same time dilation relative to each other. This leads to the paradoxical conclusion that both should be younger than the other upon reuniting.

However, this misunderstanding arises from a failure to account for the asymmetry in the situation. While it's true that each twin sees the other as moving away during the outbound and inbound journeys, the key difference lies in the fact that Alice undergoes acceleration and deceleration during her trip. At the same time, Mary remains in a single inertial frame on Earth. This change in velocity breaks the symmetry, making Alice's frame of reference different from Mary's. The acceleration phase, in particular, is crucial because it involves a shift from one inertial frame to another, which has significant implications for the passage of time.

Another common source of confusion is the assumption that time dilation only occurs during the high-speed portion of Alice's journey. In reality, time dilation affects the entire duration of her trip, including the outbound journey, the turnaround, and the return trip. Each phase of the journey contributes to the overall difference in the twins' ages, with the cumulative effect being that Alice ages far less than Mary.

It's also important to recognize that the Twin Paradox does not imply any contradiction within the theory of relativity. Instead, it highlights the counterintuitive nature of time dilation and the relativity of simultaneity. These concepts challenge our everyday experiences but are entirely consistent with the principles of Special

Relativity. The paradox arises not from a flaw in the theory but from how it forces us to rethink our assumptions about time and motion.

To further clarify these points, consider a more detailed analysis of the paradox using spacetime diagrams or the Lorentz transformations. These tools provide a mathematical framework for understanding how time dilation works, offering a clearer picture of why the traveling twin ages more slowly. By carefully examining the geometry of spacetime, we can see that the paradox is not a contradiction but a natural consequence of the relativistic nature of time.

Key Questions Raised: What Makes the Twin Paradox a Compelling Puzzle in Relativity?

The Twin Paradox is more than just an interesting thought experiment—it's a puzzle that strikes at the heart of some of the most fundamental questions in physics. It challenges our understanding of time, space, and motion, raising fundamental questions that have profound implications for our view of the universe.

One of the most compelling questions raised by the Twin Paradox is: **What is the nature of time?** In the everyday world, we think of time as something that flows uniformly, independent of our actions or movements. The paradox shatters this notion, suggesting that time can stretch or compress depending on the observer's velocity. As Einstein's theory suggests, this leads us to ask whether time is genuinely a separate, independent dimension or inextricably linked with space. If time is relative, how does this affect our understanding of events, causality, and the sequence of occurrences in the universe?

Another question the paradox raises is: **What is the true nature of simultaneity?** In Newtonian physics, two events that co-occur in one frame of reference are also simultaneous in all other frames. However, Special Relativity introduces the concept of relative simultaneity, where events in one frame may not be simultaneous in another. This leads to the realization that the order of events can differ depending on the observer's frame of reference, challenging the idea of a universal present. The Twin Paradox forces us to confront this unsettling idea, asking us to consider what it means for two people to experience time differently and whether there can ever be an objective, universal measure of time.

The paradox also raises practical questions about human experience and exploration limits. **What would it be like to travel at relativistic speeds?** If we could build spacecraft capable of approaching the speed of light, what would this mean for astronauts and their families back on Earth? Would space travel at such speeds effectively allow us to travel into the future, given the dramatic time dilation that would occur? The Twin Paradox offers a glimpse into a future where human beings might one day grapple with these questions theoretically and in practice.

Furthermore, the paradox prompts us to consider the broader implications of relativity for our understanding of the universe. **How does the relativistic nature of time and space affect the structure of the cosmos?** If time can dilate and space can contract, what does this mean for our understanding of the universe's origins, large-scale structure, and ultimate fate? The Twin Paradox is a microcosm of these more significant questions, offering a window into the mysteries of the cosmos that continue to captivate physicists and cosmologists alike.

Finally, the Twin Paradox raises philosophical questions about the nature of reality. **If time is relative, what does this mean for our experience of life, memory, and existence?** Does the paradox suggest that different observers live in other realities, each shaped by their unique experiences of time and space? These questions go beyond physics, touching on consciousness, perception, and the human condition. The paradox invites us to reflect on what it means to be human in a universe where time is not an absolute but a relative, fluid entity.

Conclusion

The Twin Paradox is a gateway into Special Relativity's strange and fascinating world. It challenges our most basic assumptions about time and space, forcing us to reconsider what we know — or think we know — about the universe. The classic scenario of the twin experiment, with its startling implications for aging and the passage of time, powerfully illustrates the counterintuitive nature of relativity.

As we continue our journey through this book, we will delve deeper into the paradox, exploring its implications, resolving its mysteries, and uncovering the profound truths it reveals about the fabric of reality. The Twin

Paradox is not just a puzzle to be solved; it is a doorway into a deeper understanding of the universe — a universe where time and space are not fixed but dynamic, interconnected, and endlessly fascinating.

4: The Mechanics of Time and Space-Time

The Concept of Space-Time: How Space and Time Are Intertwined in Relativity

The idea of space-time is one of the most profound insights to emerge from Einstein's theory of Special Relativity. Before Einstein, time and space were considered distinct and independent entities, and time ticked away uniformly for everyone, everywhere, while space extended infinitely in all directions. With its rigid, absolute structure, this Newtonian view of the universe seemed to fit with everyday experience. But as we have seen, the Twin Paradox and other relativistic phenomena challenge this notion, revealing a more intricate and interwoven relationship between time and space.

Einstein's genius lay in his ability to see beyond the apparent simplicity of Newtonian mechanics and recognize that time and space are not separate entities but two aspects of a single, unified framework—space-time. This realization came as a result of his work on the theory of Special Relativity, which showed that the speed of light is constant for all observers, regardless of their motion. To accommodate this constant speed of light, Einstein proposed that time and space must adjust accordingly, leading to space-time.

This new framework treats time as a fourth dimension, on par with the three spatial dimensions. Just as we can move forward, backward, left, right, up, and down in space, we also "move" through time. However, unlike the spatial dimensions, where movement is usually under our control, our movement through time is relentless and unidirectional—always forward, never backward, emphasizing the inevitability and power of time.

The intertwining of space and time becomes evident when we consider events in the universe. In classical physics, an event—a point in space at a specific time—would have distinct spatial and temporal coordinates. However, in the relativistic view, these coordinates are not independent. A change in an object's velocity can alter its time experience; conversely, the passage of time can influence the perception of distance. This interplay makes space-time such a powerful and essential concept in modern physics.

One way to visualize space-time is through space-time diagrams, where time is typically represented on the vertical axis and space on the horizontal axis. Each point on this diagram corresponds to an event in space-time. The worldline of an object—a curve on the chart—represents its path through space-time. This worldline is a vertical line for an object at rest, indicating it moves only through time. The worldline tilts for an object in motion, showing it moves through time and space. The slope of this line is determined by the object's velocity, with light having a unique 45-degree slope since it moves at the maximum possible speed.

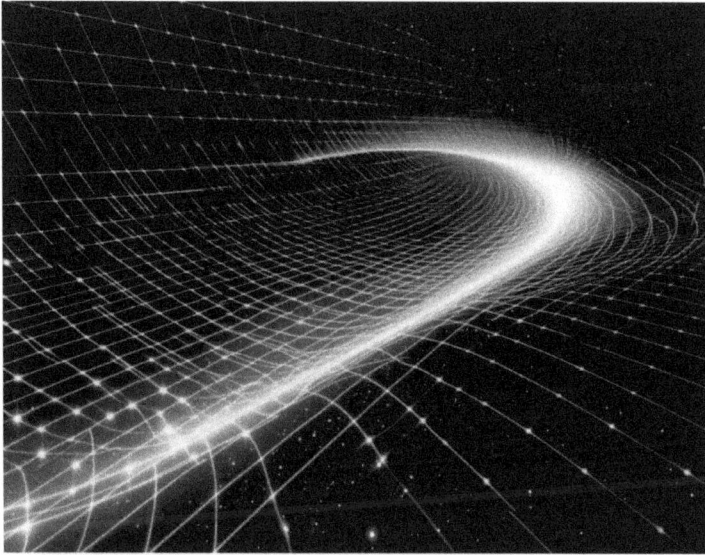

These diagrams are not just theoretical constructs; they offer real insights into the universe's operation. By plotting the worldlines of different objects, we can see how their experiences of time and space differ depending on their relative velocities. The Twin Paradox, for instance, can be illustrated in a space-time diagram, with the twin who travels at near-light speed having a worldline different from the twin who remains on Earth. The difference in the lengths of their worldlines — when measured in terms of proper time — explains why the traveling twin ages more slowly.

Understanding space-time as a unified fabric that combines space and time helps us make sense of many of the strange phenomena predicted by relativity. It shows us that time is not a separate entity but is deeply connected to space. This realization paves the way for a deeper exploration of the space-time continuum, where the true nature of our universe begins to unfold.

The Space-Time Continuum: Understanding the Four-Dimensional Fabric of the Universe

When we speak of the space-time continuum, we refer to the idea that space and time form a continuous, four-dimensional fabric that underpins the entire universe. This fabric cannot be directly observed, yet it governs everything from the motion of planets to the behavior of light.

In classical physics, space was seen as a passive stage where events played out, while time was a separate, ever-ticking clock that kept events in order. Einstein's theory of Special Relativity dismantled this view, revealing that space and time are interdependent. The space-time continuum is a more accurate representation of the universe, where events are defined not just by their position in space but by their position in both space and time.

To grasp the concept of the space-time continuum, it helps to think about it as a four-dimensional grid, where any event in the universe can be pinpointed by four coordinates: three spatial (x, y, z) and one temporal (t). This grid isn't rigid; it can be warped and stretched by mass and energy, as Einstein's later theory of General Relativity describes. However, even in Special Relativity, where we assume a flat, unwarped space-time, the idea of this continuum is crucial for understanding the behavior of objects moving at high speeds.

One of the critical features of the space-time continuum is that it allows for a new understanding of motion and rest. In Newtonian mechanics, an object at rest is not moving in space, while an object in motion changes its position over time. But in the space-time continuum, even an object "at rest" is still moving — through time. This continuous movement through time means that every object has a trajectory or worldline in the space-time continuum, and this worldline can change based on the object's velocity.

A helpful analogy to understand the space-time continuum is the surface of a calm pond. Imagine dropping a pebble into the water. The ripples spreading out from the point of impact are akin to events spreading through space-time. Now, imagine that the pond's surface is not flat but gently curved or twisted. This curvature represents how the presence of mass and energy warps the space-time continuum, affecting the paths of objects moving through it.

In the context of Special Relativity, the space-time continuum is usually considered flat, meaning that it is not warped by gravity. In this flat space-time, the effects we observe—such as time dilation and length contraction—are purely due to the relative motion of objects. However, even in this flat space-time, the continuum is not uniform. Different observers moving at different velocities will experience time and space differently. This leads to the realization that there is no single "correct" way to measure time and space; instead, these measurements depend on the observer's frame of reference.

One of the most fascinating implications of the space-time continuum is that it blurs the distinction between past, present, and future. In the classical view, the past is fixed, the present is a fleeting moment, and the future is yet to be determined. But in the space-time continuum, all events—past, present, and future—exist as points in space-time. This does not mean that the future is predetermined, but it does suggest that the flow of time is not as straightforward as it appears. Events that are future from one perspective may already be in the past from another, depending on the observer's motion through space-time.

Understanding the space-time continuum also shows why the Twin Paradox works the way it does. The twin who travels through space at near-light speed has a worldline that cuts through space-time differently than the twin who stays on Earth. When viewed in the context of the space-time continuum, this difference in worldlines leads to the time dilation effect, causing the traveling twin to age more slowly.

The space-time continuum is not just a theoretical construct; it is the very fabric of our universe. By understanding how objects move through this continuum, we gain a deeper appreciation of the relativity of time and space, and we begin to see the universe not as a static stage but as a dynamic, interconnected web of events.

Time Dilation in Depth: Mathematical Treatment of Time Dilation Effects on the Twins

Time dilation is one of the most striking predictions of Special Relativity, and it plays a central role in the Twin Paradox. To fully grasp how time dilation affects the twins, we must investigate the mathematics behind it. This mathematical treatment explains why the traveling twin ages more slowly and provides a precise way to calculate the difference in their ages upon reunion.

The starting point for understanding time dilation is the Lorentz factor, denoted by the Greek letter gamma (γ). The Lorentz factor is a function of velocity and is given by the equation:

$$\gamma = \frac{1}{\sqrt{1 - \frac{v^2}{c^2}}}$$

where:

- v is the relative velocity between the two observers (in this case, the traveling twin and the twin on Earth),

- c is the speed of light.

The Lorentz factor tells us how much time slows down for the moving observer relative to a stationary observer. When v is much less than c, the Lorentz factor is close to 1, meaning time dilation is negligible. However, as v approaches c, the Lorentz factor increases dramatically, indicating significant time dilation.

To apply this to the Twin Paradox, let's assume Alice, the traveling twin, moves at a velocity v that is a significant fraction of the speed of light while Bob remains stationary on Earth. If Alice's journey lasts t_A years as measured by her onboard clock, the time elapsed on Earth, t_B, can be calculated using the Lorentz factor:

$$t_B = \gamma t_A = \frac{t_A}{\sqrt{1 - \frac{v^2}{c^2}}}$$

This equation shows that the time elapsed on Earth t_B is greater than the time elapsed for Alice t_A. The faster Alice travels, the greater the time dilation effect and the more significant the difference in their ages will be when she returns.

Let's consider a specific example to illustrate this. Suppose Alice travels at 80% of the speed of light (v=0.8c) to a star system ten light-years away. To calculate the time it takes for her round trip, we first determine the Lorentz factor:

$$\gamma = \frac{1}{\sqrt{1 - \frac{(0.8c)^2}{c^2}}} = \frac{1}{\sqrt{1 - 0.64}} = \frac{1}{\sqrt{0.36}} \approx 1.667$$

For Alice, the round trip appears to take 25 years (10 years out, 10 years back, plus five years for the turnaround at the star). However, from Bob's perspective on Earth, the total time elapsed would be:

$$t_B = \gamma t_A = 1.667 \times 25 = 41.675 \text{ years}$$

Thus, when Alice returns, she is 25 years old, while Bob is nearly 42 years old — a difference of 17 years. This stark difference over time is the direct result of time dilation.

The mathematics of time dilation, which explains the Twin Paradox and predicts the effects of relativistic speeds on time, is not just a theoretical concept. It has been rigorously tested and confirmed by experiments involving high-speed particles and precise atomic clocks. In all cases, the predictions of time dilation have been borne out, providing strong validation for the theory of Special Relativity. This validation instills confidence in the scientific process and the accuracy of our understanding of time dilation.

Time dilation, a concept with profound implications, extends beyond the theoretical realm. In practical applications, such as particle accelerators, we can study particles moving at near-light speeds in detail. Moreover, in the context of space travel, it presents a fascinating prospect for future astronauts on long-duration missions at relativistic speeds. These astronauts would age less than those they leave behind on Earth, opening up exciting possibilities for the future of space exploration. The implications of time dilation for space travel are both hopeful and inspiring.

Your understanding of time dilation is crucial for grasping the broader implications of Einstein's theory of relativity. The mathematical treatment of time dilation provides a deeper understanding of how time behaves under the influence of high velocities. It reveals that time is not a fixed, universal quantity but a relative one that depends on the observer's motion through space-time. This insight is essential for understanding the Twin Paradox, and your role in this process is integral.

Conclusion

The mechanics of time and space-time are at the core of Einstein's theory of Special Relativity, and they are essential for understanding the Twin Paradox. The concept of space-time, with its intertwined dimensions of space and time, challenges our traditional views and opens up a new way of thinking about the universe. As a four-dimensional fabric, the space-time continuum shows us that time and space are not separate but part of a unified whole.

Time dilation, explored in depth through mathematical treatment, reveals the profound effects of high-speed motion on the passage of time. The Lorentz factor, a precise and accurate tool, provides a way to calculate these effects, showing why the traveling twin ages more slowly than the twin who stays behind. As we continue our exploration of the Twin Paradox, these concepts will serve as the foundation for understanding the mysteries of time, space, and the nature of reality itself, instilling in you a sense of the precision and accuracy of the scientific method.

In the next chapter, we will delve further into the implications of the Twin Paradox, examining how it challenges our understanding of causality, simultaneity, and the very nature of existence. The journey promises to be as enlightening as it is complex, offering new insights into the strange and fascinating world of relativity.

5: Analyzing the Twin Paradox

The Journey of the Traveling Twin: How Acceleration and Deceleration Affect Time Experienced by the Traveling Twin

The Twin Paradox offers one of the most compelling illustrations of time dilation, but its actual depth comes to light when we consider the journey of the traveling twin in detail. While time dilation due to constant velocity is well understood, the Twin Paradox requires us to delve into the effects of acceleration and deceleration on time. These phases of the journey often glossed over in simplified explanations, are crucial to resolving the paradox and understanding why the traveling twin ages differently from the twin who stays behind.

Imagine Alice, the adventurous twin, strapping herself into a spaceship designed to travel at a significant fraction of the speed of light. As she prepares to leave Earth, her journey is divided into several key phases: acceleration as the ship speeds up, constant velocity as she cruises through space, deceleration as she approaches her destination, and the return journey, which mirrors these phases.

During the acceleration phase, as the spaceship gradually increases its speed, Alice experiences forces similar to those felt by astronauts during a rocket launch. These forces, known as g-forces, result from the acceleration and directly influence Alice's time experience. According to Einstein's theory of General Relativity, which extends Special Relativity to include acceleration, time in a strong gravitational field — or under solid acceleration — slows down relative to an observer in a weaker field. While Special Relativity addresses the constant speed of light in inertial frames, General Relativity introduces the idea that acceleration affects the curvature of space-time and, thus, the passage of time.

As Alice accelerates, her spaceship can be thought of as being in a temporary gravitational field created by the acceleration itself. This effect called the equivalence principle, suggests that acceleration is indistinguishable from gravity. The stronger the acceleration, the more pronounced the time dilation. Therefore, Alice's clock ticks more slowly during acceleration than Bob's back on Earth.

Once Alice reaches her cruising speed — a velocity close to the speed of light — the effects of constant velocity time dilation dominate. This phase is well-explained by Special Relativity: as Alice moves at a continuous high speed, her clock continues to run slower than Bob's, who remains in a single inertial frame on Earth. This is the most straightforward part of the paradox, where the time dilation due to high velocity is the primary factor.

However, the return journey complicates things. As Alice decelerates upon approaching her destination and then accelerates back toward Earth, the effects of acceleration and deceleration again come into play. These phases contribute additional time dilation, further slowing Alice's clock compared to Bob's.

To grasp the cumulative effect, consider this: each phase of acceleration and deceleration adds to the overall time dilation experienced by Alice. By the time she returns to Earth, the sum of these effects causes the stark difference in aging between the twins. While it might be tempting to view acceleration as merely a transition between phases of constant velocity, it plays a critical role in the paradox. Without accounting for acceleration and deceleration, we would fail to explain why Alice ages less than Bob fully.

In essence, Alice's journey is not just a simple matter of moving at high speed; it's a complex interplay of acceleration, deceleration, and constant velocity, each affecting the passage of time differently. The slower ticking of her clock throughout these phases accumulates, leading to the striking time difference observed upon her return.

The Stay-at-Home Twin's Perspective: How the Stationary Twin's Frame of Reference Differs

While Alice embarks on her relativistic journey, Bob usually lives on Earth. From his perspective, he experiences time in the usual way — seconds, minutes, hours, days, and years tick by at a steady rate. But how does Bob perceive Alice's journey? How does his frame of reference differ from hers, and how does it contribute to the paradox?

For Bob, the most perplexing aspect of the paradox is how Alice's time appears to slow down as she speeds away. According to Special Relativity, from Bob's stationary frame of reference, Alice's clock ticks slower due to her high velocity. This is a direct consequence of time dilation: the faster Alice moves relative to Bob, the more time seems to slow down for her from his perspective.

However, it's essential to understand that Bob's frame of reference is not absolute. In the universe of Special Relativity, there are no privileged frames of reference — no perspective is more "correct" than another. What Bob observes is influenced by his position as much as by Alice's motion. From Bob's standpoint, he is stationary, and Alice is moving. But from Alice's perspective, it could be argued that she is stationary and that Bob and the Earth are moving away at high speed. This symmetry in their perspectives is what initially makes the paradox seem puzzling.

However, the critical difference is that Bob remains in a single inertial frame throughout the experiment while Alice undergoes acceleration and deceleration. This difference breaks the symmetry and is what ultimately resolves the paradox. Because Bob does not experience the same forces of acceleration that Alice does, his experience of time remains consistent and unaltered by the relativistic effects that Alice encounters.

From Bob's viewpoint, Alice's time dilation during her journey can be directly observed through telescopic communication or signals sent back to Earth. For instance, if Alice were to send regular signals back to Bob during her journey, Bob would notice that the intervals between these signals become longer as Alice accelerates and reaches her cruising speed. This observation is a real-time confirmation of the time dilation effect. As Alice decelerates to return, the signals appear to increase in frequency again, but by this point, the time difference between the twins has already been established.

Bob's perspective also includes a crucial aspect of relativity: the relativity of simultaneity. Events that are simultaneous for Bob are not necessarily simultaneous for Alice. This difference in the twins' perception of time contributes to the time discrepancy observed at the journey's end. From Bob's perspective, Alice's time experience is fundamentally different because her motion alters her relationship with the space-time continuum.

In summary, while Bob's time experience remains unaltered from his stationary frame of reference, his observations of Alice's journey reveal the effects of time dilation and the relativity of simultaneity. Bob's role in the paradox is to highlight the differences in how time is experienced depending on one's state of motion, serving as a counterpoint to Alice's accelerated journey through space-time.

Resolving the Paradox: How Accounting for Acceleration Reconciles the Different Aging Rates

At first glance, the Twin Paradox appears to be an irreconcilable contradiction. How can two identical twins end up with different ages simply because one took a journey through space? The answer lies in the detailed examination of the journey itself, particularly the role of acceleration and deceleration.

To resolve the paradox, we must account for the fact that Alice's journey involves changes in velocity while Bob remains in a single inertial frame. The resolution's core is that the acceleration and deceleration phases fundamentally alter Alice's time experience. These phases are not just transitions but are integral to understanding why the twins age differently.

When Alice accelerates at the start of her journey, she enters a different frame of reference: her clock slows relative to Bob's. As we've discussed, this slowing of time is due to the relativistic effects described by General Relativity, where acceleration mimics the effects of gravity, causing time to dilate. This initial phase sets the stage for the entire journey, as the time dilation experienced during acceleration continues to influence Alice's clock even when she reaches a constant velocity.

During the cruise phase, the constant velocity ensures that Alice's time remains dilated relative to Bob's. This is the most straightforward part of the journey to understand using Special Relativity: Alice is moving fast, so her time slows down. However, the effects of acceleration don't simply "reset" once she stops accelerating; they continue to contribute to the overall time dilation.

When Alice decelerates upon reaching her destination and again when she accelerates to return to Earth, the time dilation effects due to these phases compound the time difference further. Deceleration, like acceleration, affects Alice's experience of time, further slowing her clock relative to Bob's. By the time she begins her return journey, the accumulated time dilation from both the outbound and inbound phases means that far less time has passed for her than for Bob.

Upon Alice's return, the difference in aging is not merely a result of the high-speed cruise but the cumulative effect of all phases of her journey — acceleration, cruising at constant velocity, and deceleration. Each phase contributes to the overall time dilation, resulting in the younger age of the traveling twin compared to the twin who stayed on Earth.

Mathematically, this can be resolved using the relativistic equations for velocity and acceleration. The proper time experienced by Alice is given by integrating the time dilation effects over the entire journey, taking into account the varying velocities and the periods of acceleration and deceleration. The outcome is a clear, consistent explanation that fits within the framework of relativity.

Resolving the Twin Paradox reveals that it is not an actual contradiction but a demonstration of the counterintuitive nature of time under the effects of relativity. By carefully considering the role of acceleration, we reconcile the different aging rates and gain a deeper understanding of the relativistic universe. In this universe, time is not absolute but shaped by motion and the forces acting upon us.

Conclusion

The Twin Paradox is one of the most fascinating and educational puzzles in the theory of Special Relativity. By analyzing the journey of the traveling twin, the perspective of the stay-at-home twin, and the crucial role of acceleration and deceleration, we unravel why the twins age differently. The paradox, far from being an unsolvable riddle, is a window into the more profound nature of time and space.

Throughout this chapter, we've discovered that time is not a fixed, universal quantity but a concept intimately linked with motion and gravity. The traveling twin's time experience is altered by her journey through space-time, with each phase of acceleration and deceleration playing a crucial role in the outcome. The seemingly straightforward experience of the stay-at-home twin underscores the relative nature of time and the necessity of considering all aspects of the journey, enlightening us about the complexity of time.

As we move forward, we will continue to explore the implications of the Twin Paradox and what it reveals about the universe. The paradox deepens our understanding of relativity and challenges us to think about time, space, and reality in new and profound ways. The twins' journey, with all its complexities, is a testament to the elegance and power of Einstein's theory, offering insights that continue to resonate with scientists and philosophers alike.

6: Historical Perspectives and Theoretical Interpretations

Early Reactions: How Physicists First Responded to the Twin Paradox

When Einstein first introduced the theory of Special Relativity in 1905, it was met with awe and skepticism. The notion that time and space could be relative — interwoven into a single fabric that could stretch and contract depending on the observer's motion — was revolutionary. Among the many implications of this theory, the Twin Paradox quickly emerged as one of the most thought-provoking and controversial scenarios. It challenged the scientific community and the very fabric of how people understood time itself.

In the early 20th century, physicists were accustomed to Newtonian mechanics, where time was absolute and ticked uniformly across the universe. The idea that two identical twins could age differently simply because one traveled at high speed seemed absurd, almost sinful. The paradox seemed to undermine the principle of relativity, which stated that the laws of physics should be the same in all inertial frames. If time dilation was real, how could it be reconciled with the idea that each twin should view the other as the one who is moving?

Many early physicists struggled with the implications of the Twin Paradox. Some dismissed it as a mere thought experiment, a quirky consequence of the new theory that did not necessarily correspond to reality. Others attempted to resolve the paradox within the framework of classical physics, suggesting that some hidden factor — perhaps an undiscovered force or aether — might account for the differences in aging. These attempts, however, quickly ran into contradictions with the postulates of Special Relativity.

One of the most vocal early critics was the Dutch physicist Hendrik Lorentz, whose work on electromagnetism laid the groundwork for Einstein's theories. Lorentz developed what would later be known as the Lorentz transformations, mathematical equations that describe how time and space coordinates change between two observers in relative motion. Although his work supported the idea of time dilation, Lorentz was reluctant to accept the full implications of relativity.

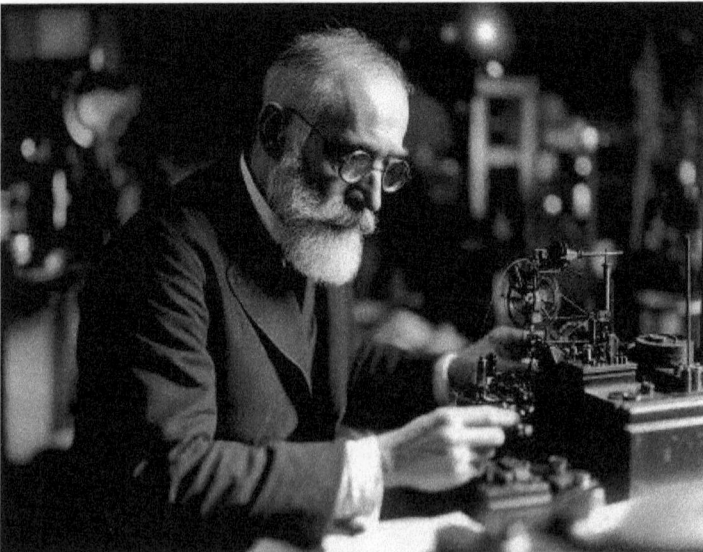

He proposed that time dilation might be an effect of motion through a stationary aether, a substance once thought to fill all space and act as a medium for light waves. However, the Michelson-Morley experiment had already cast serious doubt on the existence of aether, leaving Lorentz's interpretation on shaky ground.

As more physicists began to explore the mathematical underpinnings of Special Relativity, the Twin Paradox continued to spark debate. Some physicists, like Max Planck and Hermann Minkowski, recognized the paradox as a natural consequence of the theory and sought to understand it within the new space-time framework. Minkowski, in particular, introduced the concept of the space-time continuum, which provided a geometric interpretation of relativity and offered a clearer understanding of how time dilation and length contraction could occur.

By the 1920s, as more experimental evidence supported the predictions of Special Relativity, the Twin Paradox began to be taken more seriously. The paradox was no longer dismissed as a mere thought experiment but was recognized as an honest and profound challenge to classical notions of time and space. The early reactions to the Twin Paradox reflect the broader struggle within the scientific community to come to terms with the radical implications of Einstein's theories. This struggle would eventually lead to a deeper and more nuanced understanding of the nature of reality.

Notable Contributors: Insights from Other Key Figures in Physics

While Einstein's theory of special relativity laid the foundation for the Twin Paradox, the interpretation and understanding of the paradox were significantly enriched by the contributions of other key figures in physics. These scientists grappled with the paradox and expanded on Einstein's ideas, offering new perspectives and deepening the collective understanding of relativistic effects.

One of the most influential figures in the development of relativistic physics was Hermann Minkowski, a German mathematician whose work was crucial in shaping the modern understanding of space-time. Minkowski's 1908 lecture, in which he famously declared that "space by itself, and time by itself, are doomed to fade away into mere shadows, and only a kind of union of the two will preserve an independent reality," introduced the concept of the four-dimensional space-time continuum.

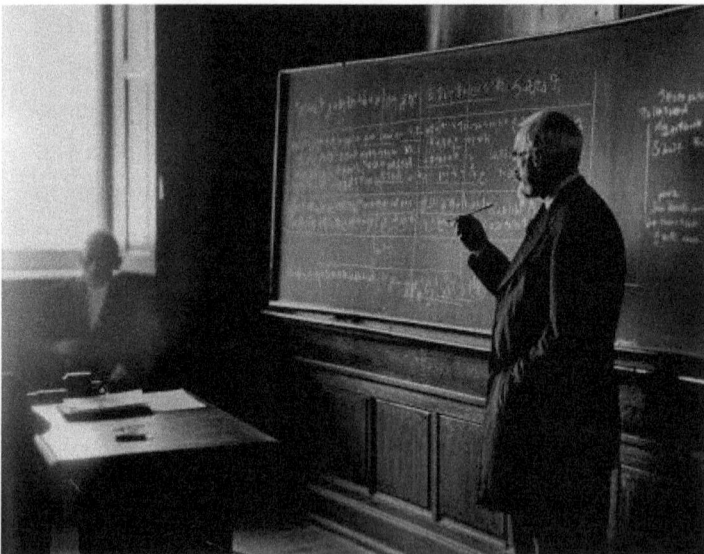

This framework allowed physicists to visualize the effects of relativity more concretely, using the geometry of space-time to explain phenomena like time dilation and the Twin Paradox. Minkowski's work showed that the paradox was not a contradiction but a natural result of how space and time are intertwined.

Langevin's work emphasized the importance of considering all phases of the journey — acceleration, constant velocity, and deceleration — in understanding the relativistic effects of time. His analysis laid the groundwork for future studies of the paradox highlighted the role of non-inertial frames of reference in Special Relativity.

Nobel laureate Max Born also significantly contributed to understanding the Twin Paradox. Born's work in the 1920s and 1930s focused on the mathematical formalism of Special Relativity, particularly the use of Lorentz transformations to describe time dilation and length contraction. Born's analyses clarified the relativistic effects experienced by the twins and reinforced the idea that the paradox could be resolved within the established framework of relativity. His work helped solidify the acceptance of the paradox as a legitimate and meaningful consequence of Einstein's theories.

American physicist Richard Feynman, known for his contributions to quantum mechanics and electrodynamics, also provided valuable insights into the Twin Paradox in the mid-20th century. Feynman was a master of simplifying complex ideas, and his lectures often included discussions of the Twin Paradox to illustrate the strange but consistent nature of relativity. Feynman's popularization of the paradox helped bring it to a broader audience, and his explanations made the counterintuitive aspects of time dilation more accessible to students and the general public.

Finally, the work of contemporary physicists like Kip Thorne and Stephen Hawking has further deepened our understanding of the Twin Paradox. In his exploration of wormholes and time travel, Thorne has used the paradox as a touchstone for discussions about the nature of time and the possibility of faster-than-light travel. In his seminal work on black holes and the nature of time, Hawking often referenced the paradox to illustrate the limits of our understanding of time and space. Both Thorne and Hawking have shown that the paradox is not just a thought experiment but a gateway to some of the most profound questions in modern physics.

These notable contributors have each brought their perspectives to the Twin Paradox, expanding on Einstein's ideas and helping to shape our modern understanding of relativistic physics. Their work illustrates how the paradox has served as a fertile ground for exploring the implications of Special Relativity, leading to new insights and more profound knowledge.

Modern Interpretations: How the Understanding of the Paradox Has Evolved

The Twin Paradox, once a source of confusion and debate, is now understood as a clear and compelling demonstration of the principles of Special Relativity. However, significant theoretical physics and experimental validation developments have marked the journey from early skepticism to modern acceptance. Our understanding of the universe and our interpretation of the paradox has grown, evolving from a puzzling thought experiment to a cornerstone of relativistic theory.

In the early days of relativity, the paradox was primarily viewed as a challenge to the new theory—an anomaly that needed to be explained or dismissed. However, as physicists began to explore the implications of time dilation and the nature of space-time, the paradox was reinterpreted as a powerful illustration of relativistic effects. By the mid-20th century, with the advent of more sophisticated mathematical tools and a better understanding of non-inertial frames of reference, the paradox was no longer seen as a contradiction but as a necessary consequence of the relativistic nature of time.

One of the key developments in the modern interpretation of the Twin Paradox has been recognizing the role of General Relativity in resolving the paradox. While Special Relativity deals with inertial frames—those moving at constant velocities—General Relativity extends the theory to include acceleration and gravity. Realizing that acceleration plays a crucial role in the paradox helped clarify why the traveling twin ages differently: it is not just the high velocity but the periods of acceleration and deceleration that contribute to the time difference. This understanding has been essential in reconciling the paradox with the broader framework of relativistic physics.

Another critical aspect of the modern interpretation is the increased emphasis on experimental validation. Over the past century, numerous experiments have confirmed the predictions of time dilation, providing empirical evidence for the effects described by the Twin Paradox. High-precision atomic clocks have been flown on aircraft and satellites, showing that clocks moving at high speeds tick more slowly than at rest. Particle accelerators have demonstrated that subatomic particles moving at near-light speeds have longer lifetimes than expected, consistent with relativistic time dilation. These experiments have confirmed the theoretical predictions and made the paradox a concrete, measurable phenomenon.

In recent decades, the Twin Paradox has also been reexamined in the context of new theoretical developments. The advent of quantum mechanics, string theory, and the search for a unified physics theory contributed to a deeper understanding of the nature of time and space. While the paradox remains a crucial example of Special Relativity, it has inspired new questions about the interplay between relativity and quantum phenomena. For instance, physicists have explored how the paradox might be understood in a universe with additional spatial dimensions, as proposed by string theory, or how it might be reconciled with the probabilistic nature of quantum mechanics.

The modern interpretation of the Twin Paradox also reflects a broader shift in our thoughts about time and space. The paradox has helped to popularize the idea that time is not an absolute, universal quantity but something that can be stretched, compressed, and experienced differently depending on the observer's state of motion. This understanding has permeated physics and popular culture, influencing everything from science fiction to philosophical discussions about the nature of reality.

Today, the Twin Paradox is more than just a puzzle to be solved—it is a fundamental example of our universe's strange and counterintuitive nature. It challenges us to think beyond our everyday experiences and consider the underlying principles that govern time and space. As our understanding of the universe continues to evolve, the Twin Paradox remains a vital and enduring part of the conversation, reminding us that even the most seemingly simple questions can lead to profound discoveries.

Conclusion

The history of the Twin Paradox is a story of how scientific ideas evolve, from initial skepticism to widespread acceptance and deeper understanding. Early reactions to the paradox reflect physicists' challenges in coming to terms with the radical implications of Special Relativity. Notable contributors like Minkowski, Langevin, and Feynman each brought new insights that helped to clarify and resolve the paradox. At the same time, modern interpretations have become a critical example of relativistic physics.

The Twin Paradox has expanded our understanding of time and space and inspired generations of physicists to explore the boundaries of our knowledge. As we delve into the mysteries of the universe, the paradox remains a powerful reminder of the elegance and complexity of Einstein's theories. This puzzle continues to provoke thought, inspire inquiry, and reveal the true nature of reality.

7: Experimental Evidence Supporting Relativity

Key Experiments: Description of Experiments Such as Atomic Clocks on Airplanes and High-Speed Particle Experiments

When Einstein first proposed his theory of Special Relativity in 1905, the ideas were radical, even revolutionary. The notion that time could slow down or space could contract depending on an observer's velocity was so counterintuitive that it demanded rigorous experimental verification. Over the past century, ingenious experiments have provided robust evidence for relativity, turning Einstein's theory from a bold conjecture into a well-established pillar of modern physics.

One of the most famous and convincing experiments supporting relativity involved using highly accurate atomic clocks. In the 1970s, American physicist Joseph Hafele and astronomer Richard Keating conducted a groundbreaking experiment involving flying atomic clocks worldwide aboard commercial airliners. Their goal was to measure the effects of time dilation—specifically, how time would slow down for clocks moving at high speeds relative to a stationary observer on the ground.

The experiment was simple in concept but complex in execution. Hafele and Keating took four cesium-beam atomic clocks, each capable of measuring time with extraordinary precision. They placed two eastward flights and two westward flights around the Earth. Upon their return, the clocks were compared to identical atomic clocks that had remained stationary at the U.S. Naval Observatory.

The results were stunning and consistent with the predictions of Special Relativity. The clocks on the eastward flight, which moved in the direction of Earth's rotation and thus traveled faster relative to the ground, showed a slight but measurable loss of time compared to the stationary clocks. The clocks on the westward flight, moving against the rotation of the Earth, gained time relative to the stationary clocks. The differences were minuscule — on the order of nanoseconds — but they were exactly what Einstein's theory predicted.

This experiment was a triumph for relativity, providing clear, empirical evidence that time dilation is real. The atomic clocks, moving at high speeds relative to an observer on the ground, experienced time differently — slowing down or speeding up depending on their direction and velocity. The Hafele-Keating experiment was not just a validation of Special Relativity; it also demonstrated the practical reality of time dilation, a concept that had once seemed purely theoretical.

Another set of experiments that provided strong evidence for relativity occurred in high-energy particle physics. In particle accelerators, such as those at CERN or Fermilab, particles are accelerated to speeds close to the speed of light. At these relativistic speeds, the effects of time dilation become profound and observable.

One such example involves the behavior of muons, subatomic particles created when cosmic rays collide with particles in Earth's atmosphere. Muons are unstable and have a very short lifespan, typically only 2.2 microseconds at rest. However, when muons are accelerated to near-light speeds in a particle accelerator, their lifespan increases dramatically due to time dilation.

In experiments conducted at particle accelerators, physicists have observed that these high-speed muons live much longer than their stationary counterparts. This extended lifespan is a direct consequence of time dilation: from the muons' perspective, time ticks more slowly due to their high velocity. The ability to measure and observe this phenomenon with such precision provides compelling evidence that the effects of relativity are tangible and measurable.

These key experiments — the Hafele-Keating atomic clock flights and the study of relativistic muons — are just a few examples of how Special Relativity has been validated through observation and experimentation. They underscore that relativity is not just a theoretical framework but a description of the natural world, with tangible, measurable effects that have been confirmed repeatedly.

Practical Applications: How Relativity Is Applied in Technologies Like GPS

Relativity is often regarded as an abstract theory with implications far removed from everyday life. However, Einstein's theories are integral to the functioning of many modern technologies, perhaps most notably the Global Positioning System (GPS). Without the corrections provided by relativity, GPS would be significantly less accurate, and the implications for navigation and communication would be profound.

GPS relies on a network of satellites orbiting the Earth, each equipped with precise atomic clocks. These satellites constantly transmit signals to receivers on the ground, which use the time stamps on these signals to calculate their position based on the time it takes for the signals to reach them. Timekeeping must be exact for this system to work with the incredible accuracy we've expected — down to the meter or even centimeter.

However, due to both Special and General Relativity, the clocks on these satellites do not tick at the same rate as clocks on Earth's surface. Special Relativity tells us that the satellites, moving at high speeds relative to the Earth's surface, will experience time dilation, causing their clocks to tick more slowly. Meanwhile, General

Relativity predicts that because the satellites are farther from the Earth's massive gravitational field, they will experience less gravitational time dilation, causing their clocks to tick faster than those on the ground.

These two effects work in opposite directions but do not cancel each other out. The net result is that the clocks on the satellites tick slightly faster than those on Earth — by about 38 microseconds per day. This might not sound like much, but even such a slight discrepancy would lead to significant errors in positioning — about 10 kilometers per day — if left uncorrected.

The GPS includes relativistic corrections in its algorithms to account for these relativistic effects. The clocks on the satellites are pre-adjusted to run slower by exactly the amount predicted by relativity, ensuring that when they are in orbit, they match the clocks on Earth. This adjustment allows the GPS to provide precise positioning data critical for everything from car navigation systems to coordinating financial transactions.

The application of relativity in GPS is a powerful example of how abstract scientific theories can have concrete, practical benefits. It demonstrates that relativity's effects are not just theoretical but have real-world consequences that must be accounted for in the design and operation of modern technology. Without Einstein's insights, GPS as we know it wouldn't work.

Other practical relativity applications are found in particle accelerators and satellite-based communication systems. In particle accelerators, relativistic effects are essential for predicting the behavior of particles at high speeds, enabling physicists to conduct experiments that probe the fundamental nature of matter. In satellite communications, corrections for time dilation are necessary to synchronize signals between ground stations and satellites, ensuring reliable global communication.

These practical applications underscore that relativity is not just a theory confined to the blackboards of physicists. A fundamental aspect of the universe subtly and profoundly affects our everyday lives.

Future Experiments: Ongoing Research and Potential Future Tests of Relativity

Even after over a century, the theory of relativity remains a rich field of exploration, with ongoing experiments designed to test its predictions with ever-greater precision. As our technology advances and our understanding of the universe deepens, new opportunities arise to probe the limits of relativity and explore its implications in previously unimaginable ways.

One area of ongoing research involves the study of gravitational waves, ripples in the fabric of space-time caused by massive objects like merging black holes or neutron stars. The detection of gravitational waves by the LIGO (Laser Interferometer Gravitational-Wave Observatory) and Virgo collaborations in 2015 provided direct evidence for a critical prediction of General Relativity. These waves were observed as tiny distortions in space-time, traveling at the speed of light, just as Einstein had predicted.

The study of gravitational waves opens up a new frontier for testing relativity. By observing these waves in more detail, scientists can test the limits of Einstein's equations under extreme conditions far beyond what can be simulated in a laboratory. Future gravitational wave observatories, such as the planned space-based LISA (Laser Interferometer Space Antenna), will allow us to detect waves from even more distant and massive sources, potentially revealing new insights into the nature of gravity, black holes, and the structure of space-time.

Another exciting area of research involves the study of time dilation at ever-increasing levels of precision. Advances in atomic clock technology have made it possible to measure time dilation effects with unprecedented accuracy. The development of optical lattice clocks, which are even more precise than the cesium atomic clocks used in the Hafele-Keating experiment, has enabled time dilation tests over much shorter distances and lower velocities. These clocks are so sensitive that they can detect the difference over time between two clocks separated by just a few centimeters in height, providing a new way to study the effects of gravity on time.

Refining these measurements may uncover subtle deviations from relativity's predictions, potentially hinting at new physics beyond Einstein's theory. Such deviations could provide clues to the long-sought-after theory of quantum gravity, which aims to unify the principles of quantum mechanics and General Relativity into a single coherent framework.

Space missions also offer new opportunities to test relativity. The proposed Deep Space Atomic Clock (DSAC) mission, which aims to place an ultra-precise atomic clock in deep space, could test the time dilation effects predicted by relativity over vast distances from Earth. By comparing the time kept by the DSAC with that of clocks on Earth, scientists could further confirm the relativistic predictions and explore potential anomalies that might suggest new physics.

Additionally, exploring black holes and studying their event horizons provide an extreme environment for testing relativity. The Event Horizon Telescope (EHT), which captured the first image of a black hole's event horizon in 2019, is a prime example of how technology enables us to test Einstein's theories in ways that were once considered science fiction. Future observations of black holes, including efforts to observe the dynamics of matter as it approaches the event horizon, could reveal new insights into the nature of space-time and test the limits of General Relativity.

In conclusion, while the theory of relativity has been confirmed through numerous experiments over the past century, the quest to understand the full implications of Einstein's insights continues. Ongoing research and future experiments promise to push the boundaries of our knowledge, testing relativity in new and extreme environments and potentially uncovering new physics that could revolutionize our understanding of the universe.

Conclusion

The experimental evidence supporting relativity is extensive and compelling, transforming what was once a bold theoretical framework into a well-validated cornerstone of modern physics. From the Hafele-Keating atomic clock experiments to the study of high-speed particles and the practical applications in GPS technology, relativity has been shown to accurately describe the behavior of time and space under a wide range of conditions.

As we look to the future, ongoing and upcoming experiments promise to test the limits of relativity further, potentially leading to discoveries that could deepen our understanding of the universe. The Twin Paradox, which once seemed like a curious thought experiment, has been validated by real-world evidence and inspires new questions and explorations. The journey of understanding relativity is far from over, and the path ahead is filled with exciting possibilities for discovery.

8: The Twin Paradox in Popular Culture

Depictions in Media: How the Twin Paradox Has Been Portrayed in Films, TV Shows, and Literature

With its mind-bending implications about time and space, the Twin Paradox has captured the imagination of scientists, writers, filmmakers, and storytellers across various media. Its inherent drama — a pair of twins, identical in every way, separated by the vastness of space and the relentless ticking of time — makes it a perfect narrative device to explore themes of identity, reality, and the human experience. Over the years, the paradox has been depicted in numerous films, TV shows, and books, each interpreting this fascinating concept.

One of the earliest and most iconic portrayals of the Twin Paradox in popular culture is Robert A. Heinlein's 1941 short story "By His Bootstraps." Although not a direct retelling of the paradox, the story deals with time travel and the circular nature of time, themes closely related to the paradox. Heinlein's work set the stage for many subsequent explorations of relativity and time travel in science fiction, blending the theoretical with the human as his characters grapple with the disorienting effects of time dilation and time's complex, non-linear nature.

In cinema, the Twin Paradox has been featured in several notable films. Christopher Nolan's *Interstellar* (2014) stands out as a compelling depiction. The film's narrative hinges on the relativistic effects of space travel, where the protagonist, Cooper, experiences time differently than his daughter back on Earth. As Cooper travels near a black hole, time dilates drastically for him, and he returns to find that decades have passed for his daughter, even though only a few years have passed for him. While the film takes some creative liberties with the science, it accurately captures the essence of the Twin Paradox, illustrating the emotional and psychological impact of time dilation on human relationships.

Another cinematic example is *Planet of the Apes* (1968), in which astronauts traveling near the speed of light return to Earth to find that thousands of years have passed and human civilization has fallen. Although the film focuses more on the shock of seeing a drastically changed world, the underlying concept is rooted in the Twin Paradox. The time dilation experienced by the astronauts makes them, in effect, time travelers, returning to a future they could never have imagined.

Television has also explored the Twin Paradox, particularly in science fiction series. In *Star Trek*, the concept of time dilation is explored in several episodes, such as "The Paradise Syndrome" from *Star Trek: The Original Series*. The show frequently uses relativistic time effects to create dramatic tension, often depicting characters who experience time differently due to their exposure to high-speed travel or proximity to intense gravitational fields. These stories entertain and introduce audiences to relativistic physics' strange and beautiful implications.

Literature, too, has had its share of explorations of the Twin Paradox. In Ursula K. Le Guin's novel *The Dispossessed* (1974), the protagonist, Shevek, experiences the effects of time dilation during his journey between planets. Le Guin uses the paradox as a plot device to explore more profound philosophical questions about time, society, and human connection. Her work illustrates how the Twin Paradox can serve as a powerful metaphor for the alienation and dislocation experienced by individuals in a rapidly changing world.

While varied in their accuracy and approach, these depictions all highlight the allure of the Twin Paradox as a storytelling tool. Whether in the form of a time-worn astronaut returning to a world that has moved on without him or a space traveler grappling with the loss of years instantly, the paradox resonates with audiences because it touches on universal themes of change, loss, and the relentless march of time.

Public Misunderstandings: Common Misconceptions Fueled by Popular Media

While the Twin Paradox has inspired countless works of fiction, its portrayal in popular media has also led to several common misconceptions. These misunderstandings often arise from the complex nature of the science involved and the need for narrative simplification, which can distort the underlying physics.

One of the most pervasive misconceptions is that time dilation is a form of time travel in the conventional sense, allowing characters to leap forward into the future at will. Films like *Interstellar* and *Planet of the Apes* contribute to this notion by dramatizing the effects of time dilation without always explaining the science behind it. As a result, many viewers come away with the impression that the Twin Paradox is a kind of "time machine" scenario rather than a natural consequence of traveling at relativistic speeds.

Another common misunderstanding is the belief that the Twin Paradox requires some form of "sci-fi magic" to work — unique technology or a mysterious force that allows time to behave differently. This misconception is often fueled by the fantastical elements accompanying time dilation in fiction, such as faster-than-light travel or wormholes. In reality, the paradox is a straightforward consequence of Einstein's theory of Special Relativity, requiring nothing more than high speeds or strong gravitational fields to manifest. However, the absence of detailed explanations in popular media can lead to the misconception that the paradox is a purely fictional or theoretical concept rather than a well-established scientific principle.

A subtler but equally important misconception is that the Twin Paradox is hypothetical and has no real-world implications. While it's true that the most dramatic examples of time dilation require speeds close to the speed of light — far beyond current human capabilities — the underlying principles are authentic and have been experimentally verified, as discussed in previous chapters. This misconception can lead to dismissing the paradox as merely a "thought experiment" rather than a window into the fundamental nature of time and space.

Popular media also sometimes portrays the effects of time dilation as reversible or inconsistent, depending on the plot's needs. Characters might experience time dilation in one direction but not the other, or they might "catch up" to lost time in ways that aren't physically possible. These portrayals can confuse audiences, leading to the mistaken belief that the Twin Paradox is a flexible concept rather than a rigorous scientific phenomenon governed by precise mathematical laws.

These misconceptions underscore the importance of accurate science communication in media. While storytelling requires a degree of artistic license, the core principles of the science being depicted must remain intact. When audiences are misled about the nature of the Twin Paradox, it distorts their understanding of relativity and diminishes the wonder and significance of the science itself.

Educational Outreach: Effective Ways to Communicate the Paradox to the General Public

Given the Twin Paradox's complexities and the misconceptions surrounding it, effective educational outreach is essential to help the general public understand this fascinating aspect of Special Relativity. The challenge lies in presenting the paradox accurately and engagingly, making the science accessible without oversimplifying or distorting the underlying concepts.

One of the most effective approaches to communicating the Twin Paradox is through interactive demonstrations and simulations. Modern technology, including computer simulations and virtual reality, allows educators to create immersive experiences that visually depict the effects of time dilation. For example, a simulation could show two twins embarking on different journeys, one staying on Earth and the other traveling near-light speeds. As the simulation progresses, viewers can see the aging process of each twin unfold in real time, illustrating the time dilation effect in a way that is both intuitive and impactful.

These visual tools can be accompanied by interactive elements that allow users to change variables such as speed, distance, and gravitational forces, helping them understand how these factors influence time dilation. By actively engaging with the science, learners can move beyond abstract concepts and develop a deeper, more personal understanding of the paradox.

Another effective outreach method is storytelling, particularly in formats that blend education with entertainment. Documentaries, for instance, can combine expert interviews, dramatic reenactments, and high-quality animations to convey the science behind the Twin Paradox in a compelling narrative format. Programs like PBS's *NOVA* or the BBC's *Horizon* have successfully used this approach to demystify complex scientific topics for a broad audience.

Storytelling can be leveraged in written formats, such as popular science books, articles, and graphic novels. By framing the Twin Paradox within a larger narrative — the story of Einstein's life, the history of physics, or a fictional tale of space exploration — authors can make the science more relatable and engaging. The key is to balance scientific accuracy with a compelling narrative, ensuring that readers come away with both a clear understanding of the paradox and an appreciation for its significance.

Public lectures and talks are another valuable tool for educational outreach. Scientists and educators can use these platforms to explain the Twin Paradox clearly and engagingly, often using analogies and real-world examples to make the concepts more accessible. TED Talks, for instance, have proven to be an effective way to reach large audiences with complex ideas, using a blend of storytelling, visuals, and clear explanations.

In addition to traditional lectures, online platforms like YouTube offer opportunities for educators to reach a global audience. Channels dedicated to science communication, such as Veritasium or Physics Girl, have successfully used short, engaging videos to explain concepts like time dilation and the Twin Paradox to millions of viewers. These videos often employ humor, visuals, and relatable scenarios to make the science more approachable, especially for younger audiences.

Finally, interactive exhibits in science museums and planetariums can provide hands-on learning experiences that bring the Twin Paradox to life. Exhibits that allow visitors to experiment with models of time dilation, explore the effects of near-light-speed travel, or visualize space-time can leave a lasting impression, making the abstract concepts of relativity tangible and memorable.

These educational outreach efforts, whether through simulations, storytelling, lectures, or exhibits, share a common goal: to make the Twin Paradox—and the science of relativity—understandable and exciting for everyone. By effectively communicating the paradox, educators can inspire curiosity and foster a deeper appreciation for the wonders of modern physics.

Conclusion

The Twin Paradox has found a unique place in popular culture, captivating audiences with its intriguing implications about time and space. However, its portrayal in media has also led to several misconceptions that can obscure the true nature of this scientific phenomenon. Through effective educational outreach—using tools like interactive simulations, storytelling, and public talks—these misconceptions can be addressed, helping to ensure that the paradox is understood not just as a captivating narrative device but as a profound and accurate reflection of the reality described by Einstein's theory of relativity.

As we continue exploring the Twin Paradox in this book, we will delve further into its implications, not just for science but for our broader understanding of the universe and our place within it. The paradox remains a powerful tool for education and inspiration, bridging the gap between the complex world of theoretical physics and people's everyday experiences worldwide.

9: Extensions and Connections to Other Theories

General Relativity and the Twin Paradox: How the Paradox Fits into Einstein's Broader Theory of General Relativity

As we've explored in previous chapters, the Twin Paradox is rooted in Einstein's Special Relativity, which deals with the behavior of objects moving at constant speeds in inertial frames of reference. However, to fully understand the implications of the paradox, we must extend our discussion to Einstein's later work—General Relativity. While Special Relativity focuses on high velocities, General Relativity brings gravity into the picture, offering a more comprehensive understanding of the universe that incorporates both time dilation due to relative motion and time dilation due to gravity.

In Special Relativity, the Twin Paradox arises from the difference in time experienced by the twin who travels at high speed compared to the twin who remains stationary. But what happens when we introduce gravity into the equation? General Relativity tells us that gravity, like high velocity, can warp time. This phenomenon is known as gravitational time dilation, where time passes more slowly in more vital gravitational fields.

To visualize this, imagine a clock placed on the surface of the Earth and another identical clock placed in a satellite orbiting the Earth. According to General Relativity, the clock on the Earth's surface, closer to the planet's massive gravitational field, ticks more slowly than the clock in orbit. This effect is not just theoretical; it has been confirmed by experiments involving atomic clocks on Earth and in orbit, similar to those used in GPS satellites, as discussed in previous chapters.

Now, let's connect this to the Twin Paradox. Suppose one twin stays on Earth while the other embarks on a journey that involves both high speeds and passage near a massive object, such as a black hole. The traveling twin would experience both types of time dilation—due to their high velocity (Special Relativity) and due to the intense gravitational field of the black hole (General Relativity). The combination of these effects would result in even more significant time dilation, with the traveling twin aging far less than the twin who remains on Earth.

This scenario is not purely hypothetical. It was depicted in Christopher Nolan's *Interstellar*, where the protagonists experience extreme time dilation near a supermassive black hole. While the movie takes creative liberties, the underlying science is grounded in General Relativity. The gravitational pull of the black hole significantly slows down time for those near it, illustrating how General Relativity can amplify the effects of the Twin Paradox.

The inclusion of gravity through General Relativity also highlights another important aspect of the paradox: the role of acceleration. In Special Relativity, acceleration is treated as a transition between inertial frames, but in General Relativity, it's seen as indistinguishable from gravity, thanks to the equivalence principle. This principle states that the effects of gravity and acceleration are locally indistinguishable, meaning that being in a gravitational field is equivalent to accelerating through space. Thus, when the traveling twin turns around to return to Earth, experiencing acceleration, this can be thought of as experiencing a form of gravity, further affecting the passage of time.

When viewed through the lens of General Relativity, the Twin Paradox becomes not just a puzzle about relative motion but a demonstration of how time is malleable, stretched, and compressed by both speed and gravity. This extension of the paradox deepens our understanding of time dilation, showing that it's not just about how fast you're moving but also where you are in the universe's gravitational landscape.

Quantum Mechanics and Time: Possible Links Between the Paradox and Quantum Theories

As we venture into the quantum realm, the Twin Paradox takes on new dimensions of complexity and intrigue. Quantum mechanics, with its principles of uncertainty, superposition, and entanglement, presents a view of the universe that often seems at odds with the deterministic nature of relativity. Yet, researchers have long sought to reconcile these two pillars of modern physics, leading to fascinating speculations about how time—so central to the Twin Paradox—might behave in the quantum world.

One of the most intriguing connections between the Twin Paradox and quantum mechanics comes from the concept of quantum time dilation. Recent studies have suggested that time dilation, as predicted by relativity, might have observable effects on quantum systems. For example, consider a particle in a superposition of states, which exists simultaneously in multiple locations or velocities. According to quantum theory, this particle would experience time differently in each state. When combined with relativistic time dilation, this could lead to a situation where the particle ages differently in each state, creating a kind of quantum Twin Paradox.

This idea has been explored in experiments with atomic clocks and quantum particles, where researchers have observed time dilation effects on particles moving at different velocities. Though still in their early stages, these experiments hint at a deeper connection between relativity and quantum mechanics. This suggests that time dilation might not be a purely classical phenomenon but extends into the quantum realm.

Another fascinating area of exploration is the concept of quantum entanglement and how it might interact with relativistic time dilation. Entanglement, often described as "spooky action at a distance," occurs when two particles become linked so that the state of one instantly influences the state of the other, regardless of the distance between them. If these entangled particles were subjected to different relativistic conditions — such as one particle moving at near-light speed while the other remains stationary — how would their entangled states evolve?

Some theorists propose that entanglement might allow for a kind of "quantum communication" that transcends the relativistic limitations of time dilation. If valid, this could open the door to new ways of thinking about information transfer and causality in the universe, potentially challenging our current understanding of quantum mechanics and relativity. However, these ideas remain speculative and are the subject of ongoing research and debate.

The intersection of the Twin Paradox and quantum mechanics also raises profound questions about the nature of time itself. In quantum theory, time is often treated as a parameter, a backdrop against which events unfold, rather than a dynamic entity that can be stretched or compressed. This contrasts sharply with relativity, where time is a central player, intertwined with space and affected by gravity and motion. Reconciling these two views of time is one of the significant challenges of modern physics, and the Twin Paradox provides a fascinating context in which to explore these questions.

Could there be a quantum theory of gravity that unifies these disparate views of time? Some physicists believe that the key to such a theory lies in understanding how time behaves at the most minor scales of the universe — at the level of quantum particles and fields. Suppose we could develop a theory that integrates the quantum and relativistic views of time. In that case, it might offer new insights into the Twin Paradox and other time-related phenomena, such as the nature of black holes, the early moments of the universe, and even the possibility of time travel.

While these ideas are still in theoretical physics, they illustrate the connections between the Twin Paradox, quantum mechanics, and our understanding of time. As research continues, we may discover that the paradox is not just a puzzle of relativity but a gateway to a deeper, more unified understanding of the universe.

Philosophical Implications: The Paradox's Implications for Our Understanding of Time, Reality, and Causality

Beyond the scientific and theoretical discussions, the Twin Paradox raises profound philosophical questions about the nature of time, reality, and causality. These questions have intrigued philosophers for centuries, long before Einstein revolutionized our understanding of the universe, and they continue to be an affluent area of exploration.

At the heart of the Twin Paradox is the idea that time is not absolute but relative, depending on the observer's state of motion. This challenges our everyday experience of time as a uniform, unchanging flow from the past through the present to the future. If time can be stretched or compressed based on speed or gravity, what does that say about the nature of reality? Is time a fundamental feature of the universe, or is it a construct that emerges from the underlying laws of physics?

Philosophers have long debated these questions, often framing them in terms of different time models. For example, the "block universe" model suggests that all moments in time — past, present, and future — exist simultaneously in a four-dimensional space-time block. In this view, time is like a landscape, with all events laid out in a static, unchanging structure. The Twin Paradox fits neatly into this model, implying that different observers can experience different slices of this landscape depending on their motion, but the overall structure remains fixed.

However, this view of time challenges our intuitive sense of free will and causality. If the future already exists, how can we influence it? Are our choices predetermined, or do we have the power to shape the future? The Twin Paradox forces us to confront these questions by illustrating how time can be manipulated. If one twin can age more slowly than the other simply by traveling at high speed, does that mean they are experiencing a different "reality"? And if so, what does that say about the nature of time itself?

Another philosophical implication of the Twin Paradox is its challenge to the concept of simultaneity. In everyday life, we assume that events occur in a single, universal present moment. But as we've seen, relativity teaches us that simultaneity is relative — what is "now" for one observer may be in the past or future for another. This has led to debates about the nature of time and whether there can be an actual, objective present moment. Some philosophers argue that the lack of absolute simultaneity undermines the concept of "now," leading to the conclusion that time may be an illusion, a construct of our perception rather than a fundamental aspect of reality.

The paradox also touches on more profound questions about causality and the nature of time travel. If time can be manipulated through high speeds or gravitational fields, as the paradox suggests, could this open the door to backward time travel? And if so, what would that mean for causality — the idea that cause precedes effect? The paradox forces us to consider scenarios where time loops back on itself or where different observers experience events in various orders. These scenarios challenge our conventional understanding of cause and effect, raising the possibility of paradoxes such as the "grandfather paradox," where a time traveler could potentially alter the past and prevent their existence.

While these philosophical implications may seem abstract, they have real consequences for understanding the universe and our place within it. The Twin Paradox invites us to question our most basic assumptions about reality by highlighting time's strange and counterintuitive nature. It reminds us that the universe is far more complex and mysterious than our everyday experiences suggest and that the nature of time, far from being a simple, linear progression, is a profound and intricate tapestry woven into the very fabric of the cosmos.

Conclusion

Though seemingly a simple thought experiment, the Twin Paradox opens up a world of connections and implications extending far beyond Special Relativity's confines. By exploring its relationship with General Relativity, quantum mechanics, and the philosophical questions of time and reality, we gain a deeper understanding of the paradox and its significance in the broader context of modern physics and philosophy.

As we continue our journey through the Twin Paradox and its implications, we are reminded that this paradox is not just a puzzle to be solved but a profound insight into the nature of the universe. It challenges us to rethink our assumptions about time, reality, and causality and invites us to explore the deepest mysteries of the cosmos. Whether through the lens of relativity, quantum mechanics, or philosophy, the Twin Paradox remains a powerful tool for understanding the universe – and our place within it.

10: Conclusion and Reflections

Summary of Key Insights: Recap of the Main Points and Resolutions Discussed in the Book

As we reach the end of our journey through the Twin Paradox, it is worth reflecting on our key insights and discoveries. This thought experiment, seemingly simple in its setup, has unraveled to reveal a profound and complex understanding of time, space, and reality itself. The Twin Paradox serves as a gateway into the broader world of Special Relativity, challenging our preconceived notions and guiding us toward a deeper comprehension of the universe.

We began by exploring Special Relativity and introducing its two core postulates: the constancy of the speed of light and the principle of relativity. These ideas shattered the classical view of absolute time and space, revealing instead that time is elastic, stretching or contracting depending on the observer's frame of reference. This was the first step in understanding how two identical twins, separated by a journey at relativistic speeds, could age at different rates.

The journey continued with a detailed analysis of time dilation and length contraction — two cornerstones of relativity that explain why the traveling twin in the paradox ages more slowly. Time dilation, the phenomenon where time slows down for objects in motion relative to a stationary observer, provided the first clue. We explored this through both theoretical explanations and experimental evidence, such as the Hafele-Keating experiment with atomic clocks on airplanes and the behavior of high-speed particles in accelerators.

We then delved into the specifics of the Twin Paradox, breaking down the journey of the traveling twin and the perspective of the stay-at-home twin. By examining how acceleration and deceleration contribute to time dilation, we resolved the apparent paradox: the traveling twin's experience of time is fundamentally different due to the relativistic effects of their journey. This resolution highlighted the importance of considering all phases of motion—not just constant velocity—in understanding the paradox.

As we expanded our exploration, we saw how the paradox fits into the broader framework of General Relativity, where gravity plays a role in warping time. We also touched upon the intriguing connections between the paradox and quantum mechanics, considering how time dilation might behave in the quantum realm. These connections opened up new avenues for thought and potential future research, suggesting that the Twin Paradox is not just a curiosity of relativity but a key to unlocking more profound mysteries of the universe.

Throughout the book, we also examined the impact of the Twin Paradox on popular culture and public understanding. From its depictions in films and literature to the common misconceptions that have arisen, we recognized the importance of effective science communication in conveying the true nature of this paradox. Educational outreach, we argued, is crucial in ensuring that the paradox is understood not just as a narrative device but as a natural, experimentally verified phenomenon.

In summary, the Twin Paradox has served as a lens through which we've explored the fascinating world of relativity. It has challenged our understanding of time and space, pushed the boundaries of theoretical physics, and inspired scientific inquiry and artistic expression. The insights we've gained from this paradox are not just about the aging of twins but about the very nature of the universe we inhabit.

The Impact of the Twin Paradox: Its Significance in Physics and Beyond

While often regarded as a thought experiment, the Twin Paradox has profoundly impacted the field of physics and beyond. Its significance lies not only in its role as a demonstration of the effects of Special Relativity but also in how it has influenced our understanding of time, reality, and the nature of the universe.

In physics, the Twin Paradox has been a powerful tool for teaching and exploring the implications of Einstein's theories. It provides a clear and tangible way to illustrate the concept of time dilation—a cornerstone of relativity that is often difficult to grasp through abstract equations alone. By framing time dilation in terms of a story about two twins, the paradox makes this complex concept accessible and relatable, helping students and physicists visualize and understand the strange but real effects of high-speed travel.

The paradox has also driven experimental research, motivating scientists to test the predictions of relativity with increasing precision. Experiments like those involving atomic clocks, high-speed particles, and GPS satellites have all been influenced by the questions posed by the Twin Paradox. These experiments have confirmed the predictions of relativity and led to technological advancements that impact our daily lives. The GPS, for instance, relies on the very principles of time dilation illustrated by the Twin Paradox, demonstrating the practical significance of this thought experiment.

Beyond physics, the Twin Paradox has permeated popular culture, inspiring countless works of science fiction, literature, and film. It has become a symbol of the broader themes of time travel, identity, and the relativity of experience—ideas that resonate deeply with human concerns about aging, memory, and the passage of time. By exploring the paradox through artistic and narrative forms, creators have brought the abstract ideas of relativity into the public consciousness, making them part of our cultural lexicon.

The philosophical implications of the Twin Paradox are also profound. The paradox challenges our understanding of time as a linear, unchanging flow, suggesting instead that time is relative and malleable. This has led to philosophical debates about the nature of reality, causality, and free will. If time can be experienced differently depending on one's motion, what does that say about the nature of existence? Are past, present, and future fixed, or are they fluid, shaped by our movement through space-time? These questions, prompted by the Twin Paradox, inspire deep reflection and discussion among philosophers and scientists alike.

The paradox's influence extends further into education and public engagement with science. It serves as a bridge between the complexities of theoretical physics and the general public's curiosity. By sparking interest and wonder, the Twin Paradox can inspire future generations of scientists, encouraging them to explore the universe's mysteries with the same curiosity that drove Einstein himself.

In essence, the Twin Paradox's impact goes far beyond the confines of a thought experiment. It has shaped our understanding of time and space, influenced scientific research and technological development, inspired cultural and philosophical exploration, and fostered a deeper public engagement with the wonders of modern physics. Its significance is a testament to a simple yet profound idea's enduring power.

Final Thoughts: Reflecting on the Journey Through the Paradox and Its Implications for Future Research

As we conclude our exploration of the Twin Paradox, it is worth reflecting on our journey and the broader implications of this thought experiment for future scientific research.

With its elegant simplicity, the Twin Paradox has provided us with a profound insight into the nature of time and space. What began as a hypothetical scenario has become a cornerstone of modern physics, challenging our intuitions and expanding our understanding of the universe. The paradox shows that time is not a fixed entity but a dynamic, relative phenomenon that depends on motion and gravity. This realization has not only deepened our knowledge of relativity but has also opened up new avenues for exploration, from quantum mechanics to cosmology.

Looking forward, the Twin Paradox continues to hold promise for future research. As we delve deeper into the connections between relativity and quantum mechanics, the paradox may serve as a guiding principle for unifying these foundational theories. For example, the interplay between time dilation and quantum entanglement could reveal new insights into the nature of time at the quantum level, potentially leading to breakthroughs in our understanding of quantum gravity and the structure of space-time.

The paradox also has implications for our exploration of the cosmos. As humanity ventures further into space, with missions to Mars and beyond, the effects of time dilation will become more than just a theoretical concern—they will be a practical reality for astronauts traveling at high speeds or in strong gravitational fields. Understanding these effects will be crucial for planning long-duration space missions and ensuring the well-being of those who embark on them.

Moreover, the Twin Paradox invites us to reconsider our philosophical and existential perspectives. It challenges us to think about time not as a uniform progression but as a variable experience shaped by our position and motion in the universe. This perspective can potentially influence science and our broader cultural and philosophical outlooks, encouraging a more nuanced understanding of reality and our place within it.

In conclusion, the Twin Paradox is more than just a puzzle to be solved — it is a window into the fundamental nature of the universe. It has taught us that the universe is far stranger and more wondrous than we might have imagined, and it has inspired us to continue questioning, exploring, and discovering. As we close this chapter, we are reminded that the journey of understanding is never complete. The paradox began with a simple question — how can two twins age differently? — has led us on a journey of discovery that continues to unfold, revealing new mysteries and possibilities with each step.

In its simplicity and profundity, the Twin Paradox will continue to inspire future generations of scientists, thinkers, and dreamers. It is a testament to the power of curiosity and imagination — a reminder that even the most abstract ideas can lead to profound insights into the nature of the universe. As we move forward, let us carry with us the lessons of the Twin Paradox, using them as a guide to explore the infinite possibilities that lie ahead.

Glossary

Acceleration:
Acceleration is the rate at which an object changes its velocity. In the context of the Twin Paradox, acceleration plays a crucial role in differentiating the experiences of the twins, mainly when one twin changes speed and direction.

Atomic Clock:

A highly accurate clock that measures time-based on the vibrations of atoms, usually cesium or rubidium. Atomic clocks, such as the Hafele-Keating experiment, are used to test the effects of time dilation.

Causality:
The principle that cause precedes effect. In relativity, the concept of causality can be challenged by scenarios involving time dilation and the relativity of simultaneity.

The constancy of the Speed of Light:

A fundamental postulate of Special Relativity states that the speed of light in a vacuum is always the same, regardless of the observer's motion or the light source. This postulate leads to the phenomena of time dilation and length contraction.

Equivalence Principle:

A key concept in General Relativity states that the effects of gravity and acceleration are locally indistinguishable. This principle explains why acceleration affects time, like gravity.

Event Horizon:

The boundary around a black hole beyond which no information or matter can escape. In the context of relativity, the event horizon is an extreme environment where the effects of time dilation become very pronounced.

General Relativity:

Einstein's theory extends Special Relativity to include gravity. It describes how mass and energy warp space-time, leading to the effects of gravitational time dilation and the curvature of space-time.

Global Positioning System (GPS):

A satellite-based navigation system that relies on precise time measurements. The accuracy of GPS depends on accounting for time dilation effects predicted by both Special and General Relativity.

Gravitational Time Dilation:

A consequence of General Relativity, where time passes more slowly in more vital gravitational fields. This effect is observed, for example, in clocks on Earth's surface compared to clocks in orbit.

Hafele-Keating Experiment:

An experiment was conducted in 1971, during which atomic clocks were flown on airplanes to test the predictions of time dilation. The results confirmed that clocks moving at high speeds experience time differently than at rest.

Inertial Frame of Reference:

A frame of reference in which an object is either at rest or moving at a constant velocity. Special Relativity primarily deals with inertial frames, while General Relativity addresses non-inertial frames involving acceleration.

Length Contraction:

A phenomenon predicted by Special Relativity where the length of an object moving at high speed appears shorter along the direction of motion when observed from a stationary frame.

Lorentz Factor (γ):

A mathematical factor describes how much time dilation, length contraction, and relativistic mass increase occur at a given velocity. The equation provides it.

$$\gamma = \frac{1}{\sqrt{1 - \frac{v^2}{c^2}}}$$

where v is the object's velocity, and c is the speed of light.

Muons:
Subatomic particles with a short lifespan are used in experiments to study time dilation. Due to time dilation effects, muons live longer when accelerated to near-light speeds.

Photon:
A quantum of light or electromagnetic radiation. Photons travel at the speed of light and are often used in discussions of relativity to illustrate the constancy of the speed of light.

Principle of Relativity:

The principle states that the laws of physics are the same in all inertial frames of reference. This principle is a cornerstone of Einstein's theory of Special Relativity.

Proper Time:

The time interval is measured by a clock at rest relative to the measured event. In the Twin Paradox, the proper time is the time experienced by the twin who remains in a single inertial frame.

Quantum Mechanics:

The branch of physics deals with phenomena at the most minor scales, such as particles and their interactions. Quantum mechanics introduces concepts like superposition and entanglement, which can intersect with relativistic ideas in discussions of quantum time dilation.

Relativity of Simultaneity:

The concept is that simultaneous events in one frame of reference may not be simultaneous in another, moving relative to the first. This idea challenges the notion of a universal present and is a crucial aspect of Special Relativity.

Special Relativity:

Einstein's theory describes the behavior of objects moving at constant speeds close to the speed of light. It introduces concepts such as time dilation, length contraction, and the constancy of the speed of light.

Space-Time:
The four-dimensional continuum combines the three spatial dimensions with time. In relativity, space and time are intertwined, and events are described by their position in space-time.

Spacetime   **Diagram:**
A graphical representation of events in space-time, where time is usually plotted on the vertical axis and space on the horizontal axis. Spacetime diagrams are used to visualize phenomena like time dilation and length contraction.

Time   **Dilation:**
A phenomenon predicted by Special Relativity where time passes more slowly for an object in motion relative to a stationary observer. This effect is central to the Twin Paradox and has been confirmed by various experiments.

Twin   **Paradox:**
A thought experiment in Special Relativity where one twin travels at near-light speed while the other remains on Earth. Upon return, the traveling twin was younger than the twin who had stayed behind due to time dilation.

Worldline:
The path an object traces through space-time represents its history as a sequence of events. An object's worldline shows its motion through time and space in a spacetime diagram.

References

Books

1. "Relativity: The Special and General Theory" by Albert Einstein

    o A classic work where Einstein explains the theories of Special and General Relativity in accessible language.

2. "Spacetime and Geometry: An Introduction to General Relativity" by Sean Carroll

    o A comprehensive General Relativity textbook covering the mathematical foundations and physical implications.

3. "Six Not-So-Easy Pieces: Einstein's Relativity, Symmetry, and Space-Time" by Richard P. Feynman

    o Feynman's physics lectures focus on the complexities of relativity and the structure of spacetime.

4. "The Elegant Universe: Superstrings, Hidden Dimensions, and the Quest for the Ultimate Theory" by Brian Greene

    o Explores the connections between relativity, quantum mechanics, and string theory in an accessible way.

5. "The Road to Reality: A Complete Guide to the Laws of the Universe" by Roger Penrose

- o A detailed and comprehensive guide to the physical laws that govern the universe, including relativity.

Journals and Research Papers

1. **"The Physics of Time Travel" by David Deutsch and Michael Lockwood**

   - o Journal of Scientific American

   - o Explores the scientific possibilities of time travel within the framework of relativity.

2. **"Experimental Basis of Special Relativity in the Photon Sector" by Clifford M. Will**

   - o A review of experiments that support Special Relativity, including those involving time dilation.

3. **"Precision Tests of the Constancy of the Speed of Light" by Holger Müller et al.**

   - o Discusses modern experiments designed to test the invariance of the speed of light.

4. **"Relativity and Cosmology: A Student's Guide to Special Relativity and Its Consequences" by Derek Raine**

   - o Cambridge University Press

   - o A comprehensive guide to the principles of Special Relativity and their applications in cosmology.

5. "Experimental Tests of Gravitational Theories" by Clifford M. Will

    o   A review of experiments that have tested the predictions of General Relativity.

Online Resources and Educational Websites

1. **Einstein Online by Max Planck Institute for Gravitational Physics**

    o   An educational website that explains Einstein's theories clearly, including Special and General Relativity.

2. **The Feynman Lectures on Physics**

    o   Feynman Lectures Website

    o   Free access to Richard Feynman's famous lectures on physics, which cover many aspects of relativity.

3. **Khan Academy: Special Relativity**

    o   A series of free lessons and videos explaining the basics of Special Relativity.

4. **MinutePhysics on YouTube**

    o   Short, animated videos that explain complex physics concepts, including the Twin Paradox, simply and engagingly.

5. **MIT OpenCourseWare: Special Relativity**

    o   MIT OCW Special Relativity

o Free course materials from MIT's Special Relativity class, including lectures, assignments, and exams.

Scientific References and Further Reading

1. **"Tests of Special Relativity" by Robert H. Good**

   o American Journal of Physics

   o An article discussing various experimental tests of Special Relativity.

2. **"Gravitational Time Dilation" by A. A. Grib and Yu. V. Pavlov**

   o A paper on the effects of gravitational fields on time relevant to the broader discussion of time dilation.

3. **"General Relativity and Gravitation" by P. A. M. Dirac**

   o A foundational text on General Relativity by one of the 20th century's leading physicists.

4. **"The Nature of Space and Time" by Stephen Hawking and Roger Penrose**

   o A series of lectures by two of the most renowned physicists discussing the intersection of relativity and quantum mechanics.

5. **"Introduction to Special Relativity" by James H. Smith**

- o A textbook that provides a clear introduction to the concepts of Special Relativity, suitable for students and general readers alike.

Dave Karpinsky, PhD, MBA, PMP, is a globally recognized consultant, executive leader, and professional author whose work bridges business transformation, strategy, and personal development. With over three decades of experience advising Fortune 500 companies, government agencies, and high-growth startups where he traveled to more than 60 countries, Dave brings a rare blend of practical insight, operational excellence, and visionary thinking to every project—and every page.

His career spans top-tier consulting firms including McKinsey & Company, Accenture, SAP, Cognizant, BearingPoint, Ernst & Young, Infosys, and IBM. He has led multi-million-dollar strategic and technology initiatives for global leaders such as Capital One, Coca-Cola, Costco, DHS/TSA, Google, HP, Janus Henderson, John Deere, Lockheed Martin, McLaren, Merck, Nike, PetSmart, QuidelOrtho, and ViaSat, as well as large-scale public sector programs for the US Government, States of Alaska, Arizona, California, Florida, and Georgia.

As the author of numerous books on project turnaround, leadership, SAP implementation, and personal mastery,

Dave is known for translating complex challenges into actionable strategies that deliver measurable impact. His writing combines analytical precision with compelling storytelling—whether he's decoding enterprise system failures or exploring the psychological dynamics of decision-making and influence.

Dave holds advanced degrees in business, technology and psychology, along with a portfolio of elite professional certifications. He is a sought-after speaker, strategist, and transformation advisor who empowers individuals and organizations to break through barriers and unlock lasting success.

Outside of his professional pursuits, Dave is an avid traveler and photographer, with a passion for astrophotography and a curated collection of high-performance and exotic cars. His global perspective, intellectual curiosity, and relentless drive to improve systems and people continue to inspire readers and clients alike.

To my constant joy and loyal hearts — you make life lighter

"The twin paradox isn't a contradiction—it's a revelation: that time is personal, and motion sculpts the way we age."
— *Dave Karpinsky*

www.ingramcontent.com/pod-product-compliance
Lightning Source LLC
Chambersburg PA
CBHW031900200326
41597CB00012B/495